The Bet

The Bet

Paul Ehrlich, Julian Simon, and Our
Gamble over Earth's Future

PAUL SABIN

Yale UNIVERSITY PRESS   NEW HAVEN & LONDON

Published with assistance from the Frederick W. Hilles Publication Fund and from the foundation established in memory of Amasa Stone Mather of the Class of 1907, Yale College.

Yale University Press books may be purchased in quantity for educational, business, or promotional use. For information, please e-mail sales.press@yale.edu (US office) or sales@yaleup.co.uk (UK office).

Designed by James J. Johnson.
Set in Scala and Meta types by Integrated Publishing Solutions.
Printed in the United States of America.

The Library of Congress has cataloged the hardcover edition as follows:

Sabin, Paul, 1970–
The bet : Paul Ehrlich, Julian Simon, and our gamble over Earth's future / Paul Sabin.
pages cm
Includes bibliographical references and index.
ISBN 978-0-300-17648-3 (hardback)
1. Environmental economics. 2. Environmental policy. 3. Ehrlich, Paul R.
4. Simon, Julian Lincoln, 1932– I. Title.
HC79.E5S213 2013
333.7—dc23
2013007845

ISBN 978-0-300-19897-3 (pbk.)

A catalogue record for this book is available from the British Library.

10 9 8 7 6 5 4 3 2 1

For my parents

Contents

Preface

On a typical winter weekend morning, our house was freezing. It was the late 1970s, and my parents had set the thermostat to the low sixties. My older brother took the *Boston Globe* sports section and settled onto the hot air vent by the kitchen refrigerator. I staked out the dining room vent to read my favorite comics, wearing a wool hat.

In a way, this book has its origins in those cold childhood mornings. In the pages that follow, I tackle a huge issue: the future of humankind on the planet. At the same time, my book also attempts to answer a lingering, and more personal, question: Why exactly was our house so cold?

I was born in March 1970, a month before the first Earth Day. The environmental sentiments of the 1970s influenced my family deeply: the push to conserve everything from cans to heat, the insistence on the evil of waste. I remember hand-me-down clothes, haircuts at home, reused paper napkins, and no television. The thermostat sent a clear message. In a world of scarce resources, we needed to consume less. The little choices of daily life reflected much larger ethical decisions about the right way to live.

I held fast to that ethic in my teens, through college, and into my professional career. I wrote a regular column for the high school paper criticizing materialism and worrying about the ozone layer. I studied history and environmental studies in college. I even met my future wife on a campus recycling truck. Later, when I was in graduate school in American history in the late 1990s, I took a break from the archives to start a nonprofit, called the Environmental Leadership Program, that brought together fellow scholars interested in environmental concerns, along with peers working in advocacy organizations, businesses, and government agencies.

By this point, my thinking had evolved. I knew what I was against—locating a hazardous power plant in a poor neighborhood, for example, or slashing the Environmental Protection Agency's enforcement budget—but it was far harder to articulate what I was for. How would a "green economy" actually work? How should we manage tradeoffs among economic growth, environmental protection, and social equity? The idea for the Environmental Leadership Program was to challenge one another with questions like these—there would be no party line. Through the rough-and-tumble of argument, I hoped, we would find compelling ways to balance competing societal goals.

When I joined the history faculty at Yale University in 2008, I wanted to keep thinking about these issues, particularly our society's inability to agree on what to do about climate change and other key problems. Writing about the rise of the environmental movement since the 1960s, and the backlash and debates it engendered, offered me a way to examine the striking divide that has emerged between liberals and conservatives on environmental questions.

Republicans and Democrats passed landmark environmental laws together in the early 1970s, but in the ensuing decades, the parties have increasingly diverged. What were the roots of this partisan divide? Scholars often explain the change by pointing to the political parties' shift to more ideologically coherent and regionally defined blocs that used the environment as a wedge issue. In this interpretation, Republicans abandoned the environment to Democrats. An alternate explanation emphasizes an economic backlash, with business groups—rightly or wrongly depending on political perspective—fighting expensive regulation and pushing politicians to oppose new rules. Last, many point to the creation of conservative think tanks and institutes starting in the mid-1970s, which organized a strategic media assault on environmental regulatory proposals favored by liberal advocates.

These explanations all have considerable historical evidence to support them. Yet they also do not take seriously the genuine clash between different viewpoints that occurred. Resistance to environmental legislation represented more than simply political and economic interest. Extreme claims by environmentalists, I argue, helped spark the backlash against the environmental movement in the United States and helped generate support for equally extreme positions taken by conservative opponents. Put another way, the political gulf that we see today on environmental issues has been mutually created. Only by tending to the substantive intellectual and historical elements of this divide—not just the political and economic dimensions—can we reduce the partisan conflict surrounding environmental policies and find a more pragmatic path forward.

The rancorous clash between the biologist Paul Ehrlich and the economist Julian Simon offers a window into this gaping political divide. Concretely, their bet was about the prices of five metals. But their wager stands for much, much more—our collective gamble on the future of humanity and the planet. The bet raises hard questions about the widely held assumption among environmentalists that we are headed inexorably for a world of scarcity and likely catastrophe. It also tests conservatives' faith that free markets and technological innovation will yield continued prosperity. By better understanding both sides of this story—by really listening to the arguments they make— I hope to encourage a different conversation, in the present, about the future.

In these partisan times, one sends a book about politics into the world with trepidation. Let me be clear: I believe that we define ourselves in part through our stewardship of the planet. At the same time, there is more than one way to live on our Earth. Where I once saw resource conservation as the only possible answer to scarcity and the limits of nature, now I understand it as a far-less-certain effort to apply ethical values in a world of constantly shifting parameters and possibilities. I still keep my thermostat set low. After studying the debates between Paul Ehrlich and Julian Simon, however, moral certainties seem more elusive.

In the journey from the heating vent in my family's house to this book, I have incurred extensive debts. The Ehrlich and Simon families have been unfailingly generous with their time and stories. I thank Paul and Anne Ehrlich, Lisa Daniel, and Sally Kellock, as well as Rita James Simon, Daniel Simon, David Simon, and Judith Simon Garrett, for meeting with me or speaking on the telephone. Naomi Kleitman, Paul and Anne Ehrlich, and Sally Kellock generously provided family photos for the book. I also am grateful for the opportunity to interview

Lincoln Caplan, Aristides Demetrios, John Harte, Donald Kennedy, Charles Michener, William Nordhaus, Stephen Schneider, John Tierney, and Daniel Weinberg.

Many friends and colleagues generously commented on the manuscript, improving it immensely, including Edward Ball, Jean Thomson Black, Lincoln Caplan, Fritz Davis, Fabian Drixler, David Engerman, John Mack Faragher, Beverly Gage, Glenda Gilmore, Matthew Jacobson, Naomi Lamoreaux, Anthony Leiserowitz, David McCormick, Steven Moss, Jeffrey Park, David Plotz, Claire Potter, Tyler Priest, Jay Turner, Chris Udry, Perry de Valpine, John Wargo, Richard White, and Donald Worster. William Cronon provided valuable early advice on the book and has deeply shaped my thinking about the relation of history to environmental politics. I also have benefited greatly from conversations with Richard Brooks, Donald Chen, Jon Christensen, William Deverell, Robin Einhorn, Gregory Eow, Seth Goldman, Jacob Hacker, Daniel Kevles, Matthew Klingle, Nancy Langston, Penn Loh, Jennifer Marlon, Steven Mufson, Dara O'Rourke, Peter Perdue, Ethan Pollock, Tom Robertson, Harry Scheiber, and Jay Winter.

Jean Thomson Black, my editor at Yale, expertly steered the book to publication. Sara Hoover helped with critical final touches on the manuscript, and Laura Jones Dooley improved it with her copy-editing. David McCormick, my agent, skillfully helped me to develop the project. Gabriel Botelho, Avinash Chak, Jerrod Dobkin, Joanna Linzer, Keira Lu, and Michael Wysolmerski provided terrific research assistance along the way. I also am deeply grateful to the archivists and librarians who assisted me at the American Academy for the Advancement of Science, American Heritage Center, Bancroft Library, Jimmy Carter Library, George Washington University, Library of Congress, National Archives, Stanford University, University of Illinois, University of Maryland, and Yale University.

I am grateful to Yale University for faculty research support, including funding from the Morse Fellowship in the Humanities, A. Whitney Griswold Fund, and Frederick W. Hilles Publication Fund.

Here at Yale, I have enjoyed and benefited from the chance to work with John Wargo, Amity Doolittle, Sara Smiley Smith, Jeffrey Park, and others to develop the undergraduate major in environmental studies. For their warm collegiality and excellent insights, I also thank my Yale colleagues, including Jean-Cristophe Agnew, Ned Blackhawk, David Blight, Daniel Botsman, Garry Brewer, Becky Conekin, Dennis Curtis, Alex Felson, Paul Freedman, Joanne Freeman, Beverly Gage, Glenda Gilmore, Jay Gitlin, Robert Harms, Karen Hébert, Jonathan Holloway, Matthew Jacobson, Ben Kiernan, Jennifer Klein, Mary Lui, Daniel Magaziner, Joseph Manning, Joanne Meyerowitz, Alan Mikhail, Steven Pincus, Stephen Pitti, William Rankin, Judith Resnik, Edward Rugemer, Marci Shore, Ronald Smith, Frank Snowden, Timothy Snyder, Adam Tooze, Francesca Trivellato, Jenifer Van Vleck, Charles Walton, John Warner, and John Witt. Laura Engelstein and George Chauncey have been wonderfully supportive chairs in the Department of History. I appreciate the friendship and good humor of Dirk Bergemann, Kishwar Rizvi, Darcy Chase, Pericles Lewis, Sheila Hayre, Paige McLean, Paul El-Fishawy, Caleb Kleppner, Ted Ruger, David Simon, Michael Sloan, Leslie Stone, David Berg, and Robin Golden. Friends and colleagues from the Environmental Leadership Program continue to inspire me. Kitty Bacon generously opened her Vermont home for a few weeks each summer, and shared her secret swimming holes and donut peaches, which we enjoyed passing along to James Sturm, Rachel Gross, and Eva and Charlotte.

I am fortunate to have a remarkably supportive and loving extended family. My parents, Margery and Jim Sabin, have shared their passion for ideas and adventure, and I'm delighted to dedicate this book to them. Their house is still freezing, but it was a wonderful place to grow up, and they deserve a gold medal for parenting. Michael and Debbie Sabin leave me in awe of their commitment to teaching and education, and my nephews and niece, Zachary, Matthew, and Elena, are a joy. My wife's family, Rick and Eileen, Lara, Matt, Carter and Ella, Jill, Joel, Harper and Trevor, and Dana and David, are incredibly supportive and fun and make me feel very lucky indeed.

Writing books together with Emily these past few years has been a surprisingly fun joint effort. I love the life we've made together—you're the surest and best bet of all. My sons, Eli and Simon, have put up with our simultaneous writing and make our house sparkle with their interest in politics and curiosity about the world. At one point while writing this book, I asked Simon, then eight years old, how we would know when the world was overpopulated. "When everything starts to run out," he said. I argue in the pages that follow that it's more complicated than that, but sometimes simple claims capture essential truths. For Eli and Simon's sake, and all the other kids out there, I hope we can lay our bets carefully to create a humane and prosperous future.

Introduction

The lanky man with short black hair and sideburns almost to his chin sat down next to late-night host Johnny Carson, for *The Tonight Show*, in early January 1970. Paul Ehrlich, a thirty-seven-year-old biology professor at Stanford, leaned forward in his seat, determined to alert his national television audience to the threat he saw imperiling humanity and Earth—the danger of overpopulation. Ehrlich had made his name two years earlier with a blockbuster jeremiad, *The Population Bomb*. "The battle to feed humanity is over," Ehrlich warned in his book, predicting that hundreds of millions of people "are going to starve to death." His first appearance on *The Tonight Show* would vault him into the national consciousness as a sober prophet of impending doom.

As Carson introduced Ehrlich to millions of ordinary Americans, a new environmentalism was dawning. President Richard M. Nixon, in his State of the Union address that same month, told Congress and the nation that the "great question of the seventies" was whether Americans would "make our peace with nature." It was three months before the first Earth Day, and Nixon was about to create the Environmental Protection Agency. Despite his grim predictions, Ehrlich proved an entertaining guest, with his sharp wit, self-confidence, and booming laugh. Carson invited Ehrlich back on his show in February and again in April. At the close of each appearance, Carson flashed on the screen the address of Zero Population Growth, the organization that Ehrlich had founded to advance his agenda of population control. Up to sixteen hundred pieces of mail per day flooded into the organization's headquarters in Los Altos, California, near Stanford. Zero Population Growth quickly grew to eighty chapters across the country.[1]

At home in Urbana, Illinois, a little-known business administration professor named Julian Simon, also thirty-seven, watched Ehrlich's performances with growing dismay and envy. Carson asked Ehrlich about the relation between population growth and the food supply. Ehrlich declared, "It's really very simple, Johnny." As populations grew, food would become scarcer. Ehrlich said it was "already too late to avoid famines that will kill millions."[2]

Yet to Julian Simon, the relation between population and food was anything but simple. The Chicago-trained economist had recently written that processed fish, soybeans, and algae could "produce enough protein to supply present and future needs, and at low cost." Rather than Ehrlich's looming famines, Simon saw an ingenious technological solution that could alleviate severe protein deficiency in many countries. Distribution posed logistical challenges. But burgeoning worldwide populations would not necessarily prompt a global food shortage, Simon thought.[3]

Yet here he was, sitting and grumbling alone in his living room while the most beloved television host in the country regarded Paul Ehrlich, as Simon later complained, with a "look of stupefied admiration."[4]

Simon and Ehrlich represented two poles in the bitter contest over the future that helped define the 1970s. Ehrlich's dire predictions underlay the era's new environmental consciousness, whereas Simon's increasing skepticism helped fuel a conservative backlash against federal regulatory expansion. Ehrlich's star continued to rise through the decade. Writing and speaking engagements poured in. He appeared on Carson's show, one of the most coveted spots in television, at least twenty times. He also wrote a regular column for the *Saturday Review* and shared his fears about starvation and population growth with concerned readers in *Playboy* and *Penthouse*. Ehrlich commented broadly on nuclear power and endangered species, immigration and race relations. He readily denounced "growthmanic economists and profit-hungry businessmen" and warned of a "coming social tidal wave" due to conflicts over limited resources.[5]

Meanwhile, Simon for years played the role of frustrated and largely ignored bystander. "What could I do? Go talk to five people?" he later asked. "Here was a guy reaching a vast audience, leading this juggernaut of environmentalist hysteria, and I felt utterly helpless." There was a certain irony behind the resentment: in the late 1960s, Simon too had argued urgently in favor of slowing population growth. He had written studies arguing that birth control programs were a "fantastic economic bargain" for countries seeking to raise incomes. He had used his marketing expertise to improve the efficiency of family planning programs. But by the time Ehrlich burst onto his TV screen in 1970, Simon had changed his mind. He no longer believed that population growth posed a problem. Rather than

Ehrlich's doomsday scenarios, Simon argued that more people meant more ideas, new technologies, and better solutions. Rather than sparking the world's crises, population growth would help resolve them. People, as Simon titled his landmark 1981 tome, were *The Ultimate Resource.*[6]

The celebrity environmentalist and the little-known skeptic collided directly at the end of the 1970s, ending the decade locked in a bet that would leave their legacies forever intertwined. In 1980, Simon challenged Ehrlich in *Social Science Quarterly* to a contest that directly tested their competing visions of the future, one apocalyptic and fearful of human excess, the other optimistic and bullish about human progress.

Ehrlich agreed to bet Simon that the cost of chromium, copper, nickel, tin, and tungsten would increase in the next decade. It was a simple thousand-dollar wager: five industrial metals, ten years, prices up or down. At the same time, the bet stood for much more. Ehrlich thought rising metal prices would prove that population growth caused resource scarcity, bolstering his call for government-led population control and for limits on resource consumption. Ehrlich's conviction reflected a more general sense after the 1973 Arab oil embargo that the world risked running out of vital resources and faced hard limits to growth. Simon argued that markets and new technologies would drive prices down, proving that society did not face resource constraints and that human welfare was on a path of steady improvement. The outcome of the bet would either provide ammunition for Ehrlich's campaign against population growth and environmental calamity or promote Simon's optimism about human resourcefulness through new technologies and market forces.

Ehrlich and Simon laid their wager at a pivot point in the struggle between liberalism and conservatism in the late twentieth-century United States. With markers laid down in the pages of academic journals, their bet resonated with the cultural clash occurring in the country as a whole. The bet also captured the starkly different paths of Democrat Jimmy Carter and his Republican challenger Ronald Reagan in the 1980 election.

Jimmy Carter, a government planner and nature enthusiast, embraced conservation and limits in keeping with the idea that resources were fixed. He argued that the United States needed to adjust its consumption and production to match its "rapidly shrinking resources." Carter devoted precious political capital to changing American energy policy, considering it a national strategic priority.[7]

Ronald Reagan, by contrast, ran for office on the promise of restoring America's greatness. Reagan insisted that resource limits weren't real and shouldn't constrain America's future. In his announcement of his candidacy in November 1979, Reagan denounced "estimates by unknown, unidentifiable experts who rewrite modern history . . . to convince us our high standard of living . . . is somehow selfish extravagance which we must renounce as we join in sharing scarcity." Reagan believed that the environmental laws of the 1970s hampered the nation's economic growth. Once he beat Carter and took office, he quickly postponed hundreds of new regulations and ordered agency heads to review and rescind other burdensome rules, many of them environmental.[8]

Nixon's "environmental decade" was finished. Reagan's aggressive campaign against federal regulation helped end the political bipartisanship that characterized the modern environmental movement's early successes. The Sierra Club and other advocacy organizations surged in membership as they denounced Reagan and sought to drive his conservative appointees from office. The nation split over how cautious or bullish to be about environmental problems. The divide between liberals and conservatives and, increasingly, between Democrats and Republicans turned on the questions embedded in Paul Ehrlich's bet with Julian Simon. Did the nation and the planet face an environmental crisis? Were we running out of resources and compelled to conserve? Were there natural limits to American growth?

These questions about population growth, resources, and the fate of humanity tapped age-old intellectual traditions. Ehrlich's widely publicized fears about population growth revived the arguments of the Reverend Thomas Malthus, a political economist who famously declared in a 1798 treatise that the "power of population" exceeded "the power in the earth to produce subsistence for man." Populations doubled rapidly, Malthus argued, while subsistence could increase only incrementally. This inherent tension between populations growing exponentially and limited subsistence doomed humanity to harsh suffering. "Necessity, that all pervading law of nature," Malthus wrote darkly, kept all plants and animals to "prescribed bounds." Limited subsistence would constrain human population growth through the "grinding law of necessity, misery, and the fear of misery." Malthus's ideas about population growth, natural limits, and the struggle for existence would significantly influence Charles Darwin and Alfred Russel Wallace in their development of the theory of evolution by natural selection in the mid-nineteenth century, and be embraced by biologists like Paul Ehrlich a century later.[9]

Early critics of Malthus, however, such as the English philosopher William Godwin, anticipated Julian Simon's critique of Ehrlich, mocking Malthus's conviction that humanity was doomed to misery. Malthus's theory of relentless population growth, Godwin wrote in 1820, was just a "house of cards" that was "evidently founded upon nothing." Godwin argued that population would grow much more slowly than Malthus predicted. He also believed that humanity had barely pressed against the vast resources of the planet. Earth, Godwin wrote, could support nine billion people with little improvement in technology. Other nineteenth-century critics of Malthus, such as Friedrich Engels, thought that agricultural productivity could be "increased ad infinitum by the application of capital, labour and science." The "productive power at mankind's disposal," Engels declared, "is immeasurable." The Industrial Revolution of the nineteenth century and rapid advances in agriculture, of course, did prove Malthus wrong in the short term. The world population grew from around one billion people in 1800 to approximately three billion in 1960. But Paul Ehrlich and his contemporaries insisted that Malthus's day of reckoning had simply been deferred. Ehrlich and other new prophets of overpopulation came to be called "neo-Malthusians" for their embrace of Malthus's warnings about an inevitable gap between accelerating population growth and limited food supply.[10]

Julian Simon rejected Ehrlich's Malthusian thinking, and in doing so, Simon's views also raised venerable, even biblical, questions. What is the purpose of humans on earth? How should we measure the success of human societies? Simon was influenced by the utilitarian philosophy of Jeremy Bentham, the British philosopher. Bentham proposed that the "measure of right and wrong" in society should be "the greatest happiness of the greatest number." Following this logic, Julian Simon welcomed continued population growth because it meant that more people could live productive and meaningful lives. Bentham also had argued that "two sovereign masters, *pain* and *pleasure*," governed mankind, and he defined the good as that which maximizes pleasure and minimizes pain. Simon did not speak in the elementary terms of "pain and pleasure." But he also placed human welfare at the center of his moral universe. Simon measured societal progress in terms of human life expectancy, prevalence of disease, available food and work, and per capita income. Paul Ehrlich rejected these simple calculations of societal success. Humanity, Ehrlich thought, could not serve as the measure of all things. Humans needed to accept their proper role in a larger balance of nature on earth. Ehrlich also dismissed Simon's optimistic projections and warned that humanity's ultimate suffering would be even greater if it continued on its same path.[11]

Paul Ehrlich and Julian Simon's conflict thus continued long unresolved debates. The structure of their bet, however, matched their times. With its promise of a winner and a loser determined by the cold, hard math of natural resource prices, the bet epitomized the increasingly polarized rhetoric of American politics. Rather than sober and nuanced assessment of policy alternatives, politicians and commentators simplified complex issues and ratcheted up their opposing claims. Important insights from biology and from economics frequently were placed in opposition, without sufficient effort to reconcile their tensions and integrate them into a coherent whole. Overly grandiose claims about the constraints of nature or the power of the market fed this clash. Underlying differences in social values and attitudes toward societal risk also often were left unacknowledged. Though ritually satisfying and motivating for partisans, the rhetorical conflict helped produce legislative paralysis and deepening political rancor. Increasingly prominent political debates over climate change, for example, starting in the 1990s slipped into rhetorical ruts established in earlier debates over population growth and resource scarcity, such as the fight between Ehrlich and Simon. In this polarizing legacy, climate change became either a myth or the possible end of human civilization. Is there another way to think about the future? Instead of reading Paul Ehrlich's clash with Julian Simon as a simple white hat–black hat morality tale, their story can move us beyond stereotyped portrayals of environmentalists and conservatives. Both men, in fact, had well-considered, significant, yet competing viewpoints underlying their strong rhetoric. Ultimately the history of their bet contains cautionary lessons for both sides, and perhaps a path to a less heated, but more productive and even hopeful, conversation about the future.

CHAPTER ONE **Biologist to the Rescue**

I t was the winter of 1968, and David Brower wanted to recruit Paul Ehrlich. The longtime executive director of the Sierra Club had heard Ehrlich on the radio predicting disaster: food shortages and famines, a deteriorating natural environment, and increased conflict on a crowded planet. Now Brower wanted the thirty-five-year-old Stanford biologist to write up his ideas as a book for a Sierra Club series of paperbacks published by Ballantine Books. Ehrlich agreed. In a fit of feverish productivity, Ehrlich collaborated closely with his wife, Anne, to write the manuscript over the next few months. He wrote the draft "as 'wild' as I could" in just a few weeks and then let friends help tone it down. *The Population Bomb*, published with Paul Ehrlich as its sole author, came out in mid-1968, in an effort, Ehrlich said, to "make the population crisis an issue in this year's elections." "I will be on the 'campaign trail' for at least the rest of this academic year," Ehrlich wrote his friend Charles Birch. Ehrlich was determined to change the way Americans thought about population issues.[1]

Ehrlich delivered *The Population Bomb* to an audience receptive to grim predictions about the future. That same year saw Robert F. Kennedy and Martin Luther King Jr. assassi-nated, riots in Washington, DC, Chicago, and Kansas City, and student rebellions in Paris and Mexico City. Meanwhile, the death toll mounted in Vietnam. To these woes, Ehrlich added his warning of "vast famines" and his call for "radical surgery" to excise the "cancer" of runaway human population growth. Ehrlich folded the crises of the late 1960s into a much larger story. Humanity had enjoyed four centuries of economic growth, Ehrlich said, but "the boom is clearly over." He urged his readers to bring every argument about social problems back to sheer numbers of people. Too many cars caused smog, but it was overpopulation that created the overabundance of vehicles. More children meant more schools and more school bond debt to pay off. In order to maintain social welfare, the birthrate needed to be brought into balance with the death rate, Ehrlich warned, or "mankind will breed itself into oblivion."[2]

As *The Population Bomb* became a best-seller, going through twenty-two reprintings in the first three years, Ehrlich emerged as a prominent national spokesman on environmental issues, inundated with speaking requests. Within the framework of overpopulation, Ehrlich also addressed broader threats from excessive consumption, pesticide use, disease, and the ecologi-cal limits that he thought constrained future food production. Many environmentalists came to view the sharp-tongued, pas-sionate Ehrlich as the "best champion we got." Ehrlich's quick humor stayed relentlessly on message. At seven o'clock one Thanksgiving morning, Ehrlich answered questions on a San Francisco television show. When a woman called to tell Ehrlich that "vegetarianism was the answer . . . I replied 'only if eating salads makes men impotent.'" "What do you call people who

use the rhythm method?" Ehrlich would joke. "Parents." Ehrlich was a skilled raconteur and a master of verbal combat, the
opposite of a stereotypical brainy researcher who struggled
to explain his work. To make sure *The Population Bomb* would
reach the widest possible audience, Ehrlich paid his twelve-
year-old daughter ten dollars to read the draft manuscript and
flag any difficult passages.[3]

Ehrlich soon had a crammed schedule of public appearances that transformed him from a scientist to a celebrity. His
speaking fee increased to a thousand dollars per lecture (adjusted for inflation, around six thousand dollars in 2013). Television and radio shows called for interviews and publishers so-
licited manuscripts. "I seem to be spending more time on radio
and TV than in bed these days," Ehrlich told a friend in August
1968. On one day in Washington, DC, that month, Ehrlich did
seven radio and television shows between 7:00 a.m. and mid-
night, plus lunch with a newspaper reporter. "The book is giving me a lot of opportunity to shoot my mouth off over the public media, and I am determined to take full advantage of it,"
Ehrlich explained. Within a year of the book's publication, Ehrlich's frenetic pace had driven him to a state of exhaustion and
poor health. His doctor ordered him to curtail his activities, but
he scarcely heeded. In 1970 alone, Ehrlich gave a hundred public lectures and appeared on two hundred radio and television
shows. Each time, he returned home from a trip to dozens of
letters from people suggesting ideas and asking him questions
or seeking advice. Paul Ehrlich had arrived where he wanted to
be, on center stage, with a large and interested public audience.
For the rest of his career, Ehrlich would spend only part of his
time in active biological research, choosing to devote much of
his prodigious energy to writing and speaking about humankind's precarious relationship with the natural world.[4]

Paul Ehrlich with his sister, Sally, 1940. Courtesy of Sally Kellock.

Paul Ehrlich grew up in suburban New Jersey at the dawn of the nuclear and chemical age and during a great wave of suburban expansion. His father, William, was a shirt salesman, and his mother, Ruth, who had graduated from the University of Pennsylvania, was a homemaker. The family moved from Philadelphia to Maplewood, New Jersey, in 1941, when Paul was nine years old and his younger sister, Sally, was four and a half. The Ehrlichs were part of a migration of Jewish families from nearby cities to the suburban town with its quiet streets and excellent school system. The family even purchased a house right across from the high school. William Ehrlich traveled frequently for work, often lugging around large sample cases. He also developed Hodgkin lymphoma in his thirties, a few years after they moved to New Jersey. Between his tiring work and the debilitating illness, which finally killed him in 1955, William left most of the childrearing to Ruth. He did not care much for Paul's early interest in insects and butterflies, but Ruth encouraged Paul to explore the outdoors. Ruth Ehrlich was tough but warm, and like her son, she "didn't suffer fools lightly." After William's death, she would return to Philadelphia to become an English and Latin teacher.[5]

As a teenager, Ehrlich roamed the fields around Maplewood, often with a butterfly net in hand, exploring the pockets of nature. He had first learned to catch and preserve butterfly specimens as a young teenager at summer camp in Vermont. He thought that they were simply "beautiful," and he loved collecting things. Specimen drawers filled with butterflies soon piled up in his bedroom. Aquariums containing tropical fish cluttered the second floor. At one point, Ehrlich started sleeping in the attic to make more space in his bedroom for his collections. One day, the heat or power went off in the house, and his mother rushed to school to get him so that he could come home to rescue his fish. At the age of fifteen, Ehrlich took the train into New York City and presented himself for employment to Charles Michener, the curator of the American Museum of Natural History's butterfly collection. Michener had little money to pay a high school student worker. So he instead rewarded the young Ehrlich with colorful tropical butterflies that were unlabeled and therefore not valuable to the museum collection.[6]

Even in high school, Ehrlich showed a precocious ability in science, including a willingness to challenge the ideas of others and a love for fieldwork. He always "very much believed in himself and his ideas." At just fifteen, in 1947, Ehrlich became a charter member of the newly established Lepidopterist Society for the study of butterflies. He was one of just a handful of members from his home state of New Jersey. The following year, Ehrlich published his first scientific field notes in the society's mimeographed *Lepidopterists' News*. Ehrlich's three-paragraph report detailed his observations of butterflies at home in Maplewood, as well as in Bethesda, Maryland, where he had spent the summer. Ehrlich had examined the eye color of more than four hundred specimens of the orange sulfur butterfly. His passion for science set him apart from his peers. "He was pretty much a loner," his mother later recalled. "After all, he had a butterfly net and he was chasing butterflies, and people ridiculed him." Ehrlich learned at a young age to follow his own muse. He developed a strong belief in his ability to understand how the world worked. He saw patterns and beauty in nature that his peers simply ignored.[7]

Suburban New Jersey proved fertile ground for breeding a young environmentalist. Maplewood and its surrounding towns were a war zone in the chemical battle against insects. Large trucks would roll down the streets spraying the pesticide DDT to kill off mosquitoes. Ehrlich found it increasingly difficult to find "food plants to feed caterpillars that weren't soused in DDT." The chemical later became an academic interest for Ehrlich. His first graduate school assistantship in 1953 focused on the development of resistance to DDT in fruit flies. Housing developers also were ripping up New Jersey's farms and rolling hills and its small country roads for suburban tract housing. Ehrlich disliked how the New Jersey landscape was changing around him. He recalled later that his environmental interest grew "when I saw the subdivisions being put over the places where I used to go collect butterflies." Ehrlich was thus part of a generation of environmentalists who would grow up in the booming suburbs. The fields, woods, and backyards attracted their families, but the construction boom and the effort to control mosquitoes and other pests also threatened suburban nature and politicized many young suburbanites like Ehrlich.[8]

Ehrlich's passion for insects and biology continued at the University of Pennsylvania, where he entered college in the fall of 1949. During one of his years, Ehrlich lived in an off-campus apartment in Philadelphia with two World War II veterans. He liked to have a good time with his friends, and although he enjoyed his studies, he later described his college years as majoring in "liquor and women." With a loud voice and booming laugh, Ehrlich held forth with strong opinions on most any topic. The future of humanity provided a favored theme. Around this time, Ehrlich read Fairfield Osborn's *Our Plundered Planet* and William Vogt's *Road to Survival,* two popular 1948 warnings about overpopulation and resource scarcity. Osborn, president of the New York Zoological Society, and Vogt, a leading ornithologist, emphasized the dependence of humanity on nature. They used the recent world war to emphasize dangers posed by resource depletion and overpopulation. "Man's conflict with nature," Osborn wrote, was a "silent war" that threatened an "ultimate disaster greater even than would follow the misuse of atomic power." Describing depleted forests and shortages of arable land and the danger of population growth, Osborn warned that "another century like the last and civilization will be facing its final crisis." Osborn called for a new humility: "The time for defiance is at an end." Humanity, the "new geologic force," must "recognize the necessity of cooperating with nature." William Vogt shared Osborn's view that overpopulation and resource depletion endangered humanity's

Paul Ehrlich with his dog, Buddy, 1951. Courtesy of Sally Kellock.

survival. "Man's destructive methods of exploitation mush-
room like the atomic cloud over Hiroshima," Vogt wrote. He
insisted that man was just another "biological creature subject
to biological laws." For the young Ehrlich, Osborn's and Vogt's
books provided ample fodder for late-night discussions with
friends. He embraced their ideas that people were like other living creatures, subject to the same laws of nature and resource
constraints, and that humanity risked a dire reckoning.[9]

Ehrlich's interest in science continued to deepen, and he
decided to pursue it as a professional career. During the summers of 1951 and 1952, after his sophomore and junior years in
college, Ehrlich worked as a field officer in the Northern Insect
Survey, posted in the Canadian Arctic and Subarctic. Following
graduation from college, Ehrlich enrolled immediately in the
graduate program in biology at the University of Kansas. He
resumed work on his earliest passion, butterflies, under the su-
pervision of his high school mentor, Charles Michener, who
had moved to Kansas from New York. In an interview, Michener remembered Ehrlich fondly as a "noisy and brash" young
man who also was "bright and very capable." If you wanted to
find Michener, people would say, go listen for Paul Ehrlich's
booming voice down the hall. In this community of scientific
researchers, Ehrlich stood out as charismatic and gregarious,
with broad-ranging interests that extended far beyond his narrow butterfly research program.[10]

While in graduate school, Ehrlich met Anne Howland, an
undergraduate French major at the university just a year
younger than Paul. Anne had read some of the same books as
Paul, and she shared his worldview about population pressures
on the land. She had grown up in Des Moines, Iowa, in an artistic and literary family, with a cosmopolitan outlook. One of
her grandmothers had marched as a suffragist, and both her

Paul and Anne Ehrlich, ca. 1956. Courtesy of Paul and Anne Ehrlich.

mother and her aunt had been determined to have careers. Anne's mother wrote for the society page of the *Des Moines Register*, while her aunt worked for a prominent Chicago advertising agency. Anne's own education and career, however, was initially derailed by her relationship with Paul. The two started dating early in her junior year, and they married just a few months later, in December 1954. Their only daughter, Lisa, came unexpectedly the following November. Short on money and with a new baby in the house, Anne dropped out of college to fulfill traditional roles of wife and mother. They did not feel that they could afford a second child, and by the time that they could, it no longer seemed like the right thing to do.[11]

Despite her strong intellectual interests, Anne never finished her degree. Anne would work closely with Paul, first tapping her artistic skills to illustrate his doctoral thesis and then contributing hundreds of drawings to their 1961 book, *How to Know the Butterflies*. Drawing was work that Anne could do in spurts with a young child wandering around the house. Later, when Paul started writing essays about population, Anne became a close writing collaborator and, finally, a public figure in her own right. Paul was the "mouth" and Anne was the "brains," he liked to joke about their unusually close and symbiotic working partnership. He was the extrovert who loved being around people and making them laugh, while Anne, particularly early in their marriage, tended to recede into the background.[12]

After a postdoctoral fellowship in Chicago, Paul and Anne Ehrlich moved with Lisa to Palo Alto in 1959, where Paul began a more than fifty-year career teaching in the biology department at Stanford University. His research in population biology would result in hundreds of scientific publications, including an influential 1965 paper, coauthored with Stanford colleague Peter Raven, that helped launch the study of co-evolution: the idea that animals and plants coevolve in a series of adaptive defenses and responses. The strength of Ehrlich's analysis, and the new ecological science that he would come to represent, lay in the ways that he drew attention to the interconnectedness of humans and the natural environment. Biologists like Ehrlich and Raven showed how ecosystems functioned, and they documented rapidly changing environments, threats to endangered species, and the migration of toxic chemicals like DDT through the global food chain.[13]

As Ehrlich developed a successful research program at Stanford, he fit into a generation of biologists after World War II who saw their work in a broader political context. The furor over Rachel Carson's 1962 book, *Silent Spring*, about the dangers of chemical pesticides, helped coalesce public concerns into a new political environmentalism. Ehrlich embraced environmentalism as a secular religion. He found purpose in fulfilling humanity's moral obligation to preserve nature and restore balance between society and ecosystems. Ehrlich later described his political development as a "natural progression." "I didn't stand up one day and say 'My God, I'm going to get everybody to stop [fuck]ing.' It's sort of one thing led to another." Like many Americans in the 1960s, Ehrlich increasingly questioned whether governments and businesses could be trusted to respond effectively to the ecological crisis, when they did not disclose critical information about the risks of new chemicals and nuclear radiation. He argued passionately that Americans needed to value a simpler life, forgoing risky technologies and wasteful consumption.[14]

The Indian food crisis of the mid-1960s galvanized Ehrlich and pushed him from field research to the public stage. During the 1965–1966 academic year, Paul and Anne moved to Australia, where Paul had a National Science Foundation fellowship to study for the year. At the close of the sabbatical, Ehrlich and his family spent a few weeks in India, touring the country, as part of a longer trip that included visits to Thailand and Cambodia. At the time, news reports warned that "mass starvation" due to drought and low grain production threatened the lives of millions of Indians. From Kashmir, where Paul had hoped to obtain butterfly specimens but found the area "overgrazed," the Ehrlichs traveled to Delhi. The tremendous poverty and crowds in Delhi overwhelmed his senses and deeply shaped his thinking. "I have understood the population explosion intellectually for a long time," Ehrlich wrote about the trip to India in the opening pages of *The Population Bomb*. "I came to understand it emotionally one stinking hot night in Delhi a couple of years ago."

> My wife and daughter and I were returning to our hotel in an ancient taxi. The seats were hopping with fleas. The only functional gear was third. As we crawled through the city, we entered a crowded slum area. The temperature was well over 100, and the air was a haze of dust and smoke. The streets seemed alive with people. People eating, people watching, people sleeping. People visiting, arguing, and screaming. People thrusting their hands through the taxi window, begging. People defecating and urinating. People clinging to buses. People herding animals. People, people, people, people. As we moved slowly through the mob, hand horn squawking, the dust, noise, heat, and cooking fires gave the scene a hellish aspect.

Frightened about whether they would make it back to their hotel, Ehrlich admitted that he was simply "unaccustomed to the sights and sounds of India." But the sensory overload of the Delhi streets also gave him the overwhelmingly chaotic "feel of overpopulation." Ehrlich's revulsion at India's street life was common for Western visitors. Yet his instinct to blame sheer numbers of people, rather than their state of culture or governance, represented a shift in emphasis underway in Western thinking.[15]

Ehrlich thought that the pressure of growing populations on resources would overwhelm any advances in technology. Agricultural innovation would meet natural limits that humans could not overcome. Ehrlich criticized "narrow-minded colleagues who are proposing idiotic panaceas to solve the food problem." With funding from private foundations and the government, agronomists like Norman Borlaug were seeking ways to use fertilizer, pesticides, and new crop strains to dramatically expand agricultural output, an effort called the Green Revolution. Borlaug, who received the 1970 Nobel Peace Prize for his efforts, successfully developed a new high-yield, disease-resistant wheat strain that boosted grain production in India, alleviating the food crisis. Ehrlich thought that boosting food output at best simply bought a short amount of time for societies to take action on population growth. At worst, the Green Revolution risked aggravating the situation by ensuring that "we have an even bigger population when the crash comes." Writing to agricultural economist Lester Brown in January 1970, Ehrlich argued that "biological factors," such as growing resistance to pesticides, would bring the Green Revolution "to a halt in the not-too-distant future." Yet he noted with distress that the Green Revolution's success had persuaded many that "the population problem is over" and "any number of people" could be fed in the future.[16]

Ehrlich's fear of overpopulation and famine reflected broader elite concerns in the mid-1960s. With the world population growing from 2.5 billion people in 1950 to 3.35 billion in 1965, many commentators questioned whether the planet could sustain the growing number of people. The *New Republic* announced in 1965 that "world population has passed food supply. The famine has started." The magazine predicted that even "dramatic counter-measures" could not reverse the situation. A "world calamity" would strike within the decade. World hunger, the magazine editors wrote, would be the "single most important fact in the final third of the 20th Century." US Ambassador to India Chester Bowles concurred, telling a Senate subcommittee in June 1965 that the approaching world famine

threatened "the most colossal catastrophe in history." Starting
in January 1968, around the time that Ehrlich was writing *The
Population Bomb*, a group calling itself the "Campaign to Check
the Population Explosion" started running full-page advertisements in the *Washington Post* and the *New York Times*. The im-
agery was apocalyptic. One advertisement showed a large stop-
watch and announced that someone "dies from starvation"
every 8.6 seconds. "World population has already outgrown
world food supply," the advertisement declared. Another advertisement showed the picture of a baby under the headline
"Threat to Peace," warning that "skyrocketing population
growth may doom the world we live in." A third pictured Earth
as a bomb about to explode—with population control the only
way to defuse the threat.[17]

The violent metaphors underscored how global population
growth intersected with the Cold War struggle between the
United States and the Soviet Union. As the United States and
the Soviet Union battled for strategic advantage, they increasingly used economic development aid to win the allegiance of
"Third World" nations. Population growth endangered these
international development efforts. Hungry people, American
policy-makers believed, were more susceptible to communist
influence. Overpopulation and resource scarcity, a foreign aid
report stated emphatically in 1959, created "opportunities for
communist political and economic domination." Americans
looking abroad to prove the vitality of the capitalist model worried that their efforts would founder amid poverty and famine.
"No peace and no power is strong enough to stand for long
against the restless discontent of millions of human beings
who are without any hope," Lyndon B. Johnson warned in a
1966 speech.[18]

Fears of a Malthusian calamity that would intensify global

conflicts prompted radical prescriptions. In their 1967 book, *Famine 1975!* the brothers William and Paul Paddock argued that the United States should apply the concept of military triage to its international food aid. Countries should be separated according to whether they "can't be saved" (Haiti, Egypt, India), were "walking wounded" (Libya, The Gambia), or "should receive food" (Pakistan, Tunisia). The Paddocks' ideas of triage and limits to food aid resonated in Washington, DC. President Johnson had refused to send American wheat to India in 1966 until that country adopted a vigorous family planning program. According to presidential adviser Joseph Califano, Johnson told him, "I'm not going to piss away foreign aid in nations where they refuse to deal with their own population problems." How much Johnson and other American policy-makers believed that India faced a Malthusian crisis—and how much they needed to use the idea of a famine to sell Congress on continuing the Food for Peace export program—is a matter of historical argument. The massive scale of the eventual American relief effort is indisputable: over a two-year period, roughly one-quarter of annual US wheat production was sent to India.[19]

The Indian government denied the "famine" label and insisted that any short-term domestic food production shortages resulted from a national strategic focus on boosting industrial production. But Ehrlich, the Paddock brothers, and many others in the United States believed that population growth in India had outstripped the capacity of the nation to produce food. Ehrlich endorsed the Paddocks' call to tie food aid to population control, describing it as "coercion in a good cause." "I have been citing the Paddocks' work in all my recent talks on the population crisis," Ehrlich told a colleague in late 1967. Ehrlich described himself as "astounded" by Americans who were "horrified" by the idea of triage as applied to international food aid. The United States should announce that it would no longer ship food to countries such as India where "dispassionate analysis indicates that the unbalance between food and population is hopeless," Ehrlich declared in the British magazine *New Scientist.* "Any aid to an underdeveloped country which does not include population control aid is totally wasted," he wrote in a letter to his Australian colleague Charles Birch. These ideas did horrify many readers and listeners, such as the physicist Alvin Weinberg, director of the Oak Ridge National Laboratory. Weinberg denounced the "elitism" of wealthy and well-fed Americans declaring that the "Malthusian vise" was the only strategy to force people to have fewer babies. "When people starve, do we have *any* alternative but to try to give them food?" The idea that "misery imposed today . . . is necessary for the long-term good of man," Weinberg wrote, "is scientific arrogance at its most outrageous."[20]

Ehrlich's extreme reaction to food shortages abroad, including his willingness to cut countries off from further aid, emerged directly from his scientific research. Like many biologists, he viewed humans mostly as just another animal species. His fears about human overpopulation reflected his conclusions about the dynamics of butterflies. Butterflies existed in tenuous balance with available resources and external threats from predators and disease. No gentle "balance of nature" stabilized butterfly populations. Rather, booming and crashing population cycles characterized all animal species. Populations that grew beyond a certain threshold were brought down by resource shortages, disease, and other population-dependent factors. "The shape of the population-growth curve is one familiar to the biologist," Ehrlich wrote in a 1969 essay on human overpopulation entitled "Eco-Catastrophe." "It is the outbreak part of an outbreak-crash sequence. A population grows rapidly in the presence of abundant resources, finally runs out of food or some other necessity, and crashes to a low level or extinction." The imbalance between the birthrate and the death rate inevitably "will be redressed" in the "greatest cataclysm in the history of man." Ehrlich warned that overpopulation would lead to famine, plague, or thermonuclear war, which would raise the human death rate and reduce excessive human numbers. "We've had most of the outbreak—what remains is mainly the crash," he wrote in *Audubon* in 1970.[21]

Many other biologists shared Ehrlich's tendency to extrapolate from biological systems and nonhuman populations to the fate of human societies. Charles Remington, for example, a Yale population biologist and friend of Ehrlich who also studied butterflies, similarly connected insect populations and humans. Remington was so passionate about bugs that he once ate raw cicadas on television to show how tasty they were. In his research, Remington examined how insect populations related to plants. He found that individual insects thrived better with low population density. The lessons for human society were clear, he thought. As populations grew, tragedy would strike. At a more basic level, human lives simply would become less complete and fulfilling. "I'm scared," Remington said in a 1971 interview. "A lot of terrible things will happen to us before the population evens off. There will be famine, mass starvation, unbelievable pollution, elimination of land for recreation, education only for those who can pay for it, deterioration of services." Ecologist Eugene Odum wrote that "continued improvement of the quality of human life will be more and more threatened by cancerous growth." Humans were like parasites on the environment; humanity risked extinction by "destroying its host." To their credit, biologists like Ehrlich, Remington, and

Odum took aim at the growing sense of human control and power over nature that led to extravagant technological schemes, such as a late 1950s Project Chariot proposal to use nuclear explosions to create a deep-water port in Alaska. Yet, as Julian Simon would later argue, the biologists also showed little understanding for how people differ from butterflies and how economic systems could work to manage scarcity, drive investment and innovation, and avert shortages.[22]

The success of the US space program and pictures of the planet from space also influenced Ehrlich and his colleagues. Stunning images of Earth rising over the Moon and floating in space spurred a new way of thinking about human existence. People were alone in the universe, entirely dependent on the limited, shared, and fragile resources of the planet. "The first move must be to convince everybody to think of the earth as a space ship that can carry only so much cargo," Ehrlich said in a 1967 speech. The bleak lunar landscape also made an impression. The Moon, Mars, and Venus could not support independent human life, observed University of California, Santa Barbara, biologist Garrett Hardin. "A finite world can support only a finite population," Hardin concluded. Humans needed to learn to live within Earth's natural limits. Ehrlich, Hardin, and others called for a rapid transition to a "spaceman" economy for "Spaceship Earth." To illustrate the spaceship or life-raft concept in 1969, eighty-nine men and women in California fasted for a week and lived together in a small, two-room shelter to experience the "effects of overpopulation."[23]

As biologists like Ehrlich and Hardin captured the public imagination with the lessons of ecology, environmental problems attracted the interest of many young scientists, including two young physicists, John Holdren and John Harte. Holdren and Harte would become two of Ehrlich's closest colleagues and best friends. They later joined Ehrlich as partners in his bet with Julian Simon. Both Holdren and Harte started their academic careers in physics but soon left to pursue broader problems in environmental science. Elevated by the Cold War development of nuclear weapons and power, as well as the race into space, physicists had dominated American science funding and attention after World War II. In the late 1960s, however, disillusionment with the Vietnam War and military funding for physics research turned aspiring physicists like Holdren and Harte to more holistic approaches to societal problems. Guided by the fundamental laws of thermodynamics, in which energy cannot be created or destroyed, Holdren and Harte shared Ehrlich's faith that natural limits would constrain human society. Water, food, and energy—these fundamental building blocks of society could not be generated out of thin air. Holdren and Harte favored back-of-the-envelope calculations that could quickly show that the numbers did not add up.[24]

John Holdren met Paul Ehrlich while studying for his doctoral degree in plasma physics at Stanford. Holdren had grown up in San Mateo, California, not far from Stanford. While in high school, Holdren read the 1954 book *The Challenge of Man's Future,* by geochemist Harrison Brown. Holdren embraced Brown's interdisciplinary approach to population, resources, and technology. Elements of eugenic thinking evident in Brown's concern for "genetic soundness" and the "deterioration of the species" did not seem to get the attention of the young Holdren. Instead, Brown's book helped him to envision a career applying science to policy. After studying aeronautics and physics at MIT, Holdren enrolled in the doctoral program in physics at Stanford University. Inspired by Ehrlich's lectures and writings, John sought him out on the advice of his wife, Cheri, who studied biology in Ehrlich's laboratory.[25]

Despite the twelve-year age difference between Ehrlich and Holdren, and Holdren's status as a graduate student, the two men became close friends and collaborators. In 1969, Ehrlich and Holdren wrote an essay on population for the life sciences journal *BioScience* that said technology was not a "panacea" for population growth. Their article, developed with partial support from the Ford Foundation, criticized "Sunday-supplement conceptions of technology" that suggested "science has the situation well in hand." Farming the sea and tropics, cheap nuclear power, and irrigating desert lands could not solve the population problem. "No effort to expand the carrying capacity of the Earth can keep pace with unbridled population growth." For a biologist like Ehrlich, writing in tandem with a plasma physicist like Holdren helped bolster his credentials, particularly regarding energy and technology. Ehrlich and Holdren argued for population control over technological change to avert looming disaster. "We should ask, for example, how many vasectomies could be performed by a program funded with the 1.8 billion dollars required to build a single nuclear agro-industrial complex." After Holdren completed his doctorate in 1970 and took a position at the Lawrence Livermore National Laboratory, Ehrlich and Holdren began writing joint columns on population and environment topics for the *Saturday Review*. The essays ranged widely, including issues such as the impact of defoliants in Vietnam on food production, which Holdren and Ehrlich considered "ecocide" and "crimes against humanity."[26]

As Holdren launched his academic career in the early 1970s, he and Ehrlich continued to collaborate. "John has been working very closely with me on environmental problems," Ehrlich wrote in a 1970 letter. "Any thinking I have in the area is partly his." Ehrlich sometimes sent Holdren in his place to conferences to represent their common views. Ehrlich and Holdren edited two books on environmental issues and, along with Anne, coauthored a 1973 textbook, *Human Ecology*. Ehrlich and Holdren were fighting for what they called a "rational strategy for man." Managing population growth lay at the heart of that strategy. The two viewed population as a powerful multiplier of human impact. Together they developed an equation to represent the factors shaping humanity's environmental impact: Human Impact = Population times Affluence times Technology ($I = P{*}A{*}T$). Holdren and Ehrlich used the equation to argue against complacency about population growth's direct and negative multiplier effect on the environment.[27]

Holdren eventually joined the faculty at the University of California, Berkeley, where he cofounded the Energy and Resources Group, an interdisciplinary graduate program that he helped lead for more than twenty years. He continued to draw on his physics background to remain active on issues of nuclear arms control. Holdren played a key leadership role in the Pugwash Conferences on Science and World Affairs. In 1995, he gave the acceptance speech when that organization won the Nobel Peace Prize for its work on arms reduction and peace. Holdren, who also taught at Harvard University's John F. Kennedy School of Government, would go on to become director of the White House Office of Science and Technology Policy under President Barack Obama.[28]

John Harte, the third participant in Ehrlich's bet with Julian Simon, followed a path similar to Holdren, leaving academic physics to study environmental science. Growing up in the New York suburbs as the child of two high school teachers,

Harte shared Ehrlich's childhood fascination with suburban wildlife like frogs, snakes, and birds. He kept lists of the birds that he had seen, and his bedroom, like Ehrlich's, resembled a natural history museum. Harte's suburban nature study also developed into political environmentalism. As a teenager, Harte was angry that the suburban woods were being developed for new housing. In adolescent rebellion, he occasionally would go out at night to pull up surveyor stakes to disrupt the projects.[29]

Harte pursued a traditional academic path before being drawn into the civil rights and antiwar protests of the 1960s. Talented at math and attracted to science, Harte graduated from Harvard in 1961 and went straight to the University of Wisconsin to study physics. Following completion of his doctorate in 1965, Harte spent two years at the Lawrence Berkeley Laboratory in California. It was the height of student radicalism in the Bay Area. The previous year, during the Free Speech Movement, thousands of students had blocked the arrest of a civil rights organizer by police. Surrounding and standing on top of the police car in which the organizer was held, protesters had denounced the university's restrictive policies on campus organizing. In the months around Harte's arrival in Berkeley, antiwar protesters had burned draft cards, hung Lyndon Johnson in effigy, and sought to block trains carrying soldiers to the Oakland army base. Harte became an activist. He tutored African American students through the Congress of Racial Equality, and he became increasingly radicalized in his opposition to the Vietnam War. After moving to Yale in 1968 as an assistant professor of physics, Harte joined with other young faculty to oppose the war and protest the militarization of scientific research. On March 4, 1969, Harte, along with fellow physicist Robert Socolow, helped to shut down science classes at Yale for a day of reflection about the war and scientific connections with the military.[30]

Harte's leadership of the antiwar teach-in at Yale unexpectedly led to his transition out of physics into environmental studies. Following the event, a visiting speaker invited Harte and Socolow to participate in a National Academy of Sciences summer study project on environmental problems. They chose to study a proposed new airport in south Florida. They concluded that draining the marshes to put an airport at the edge of Everglades National Park would endanger the freshwater supplies for half a million people. Their report helped derail plans for the jetport. The experience inspired Harte to return to the Bay Area to work at the Lawrence Berkeley Laboratory's new energy and environment division, where he studied how toxic chemicals like acid rain effect ecosystems. Within a few years, Harte had started teaching an environmental problem-solving course in the Energy and Resources Group at Berkeley with John Holdren. Harte remained at Berkeley for the rest of his academic career.[31]

Harte had admired Paul Ehrlich and *The Population Bomb* from afar, but he first met Paul Ehrlich around 1974 or 1975 at a dinner with John Holdren. His first impression was that Ehrlich was "funny as hell." The two scientists became close friends. The Ehrlichs and John Holdren invited Harte to join them in the summer at the Rocky Mountain Biological Laboratory. Starting in 1961, Anne and Paul Ehrlich had spent each summer at the high-altitude biological research station, founded in 1928 in the old mining town of Gothic, Colorado. The Ehrlichs made the research laboratory a home away from home. It was a stunning research environment where they could escape on Paul's limited academic salary. Paul and Anne had first lived in an old shack left over from the mining days, cooking on a

Paul and Anne Ehrlich's cabin at Rocky Mountain Biological Laboratory, with Gothic
Mountain in the distance, July 2011. Photograph by the author.

Coleman camping stove. Later they built a one-room cabin.
Nestled in an aspen grove nine thousand feet above sea level,
the Ehrlichs' cabin offered a stunning view of nearby Gothic
Mountain, a rocky peak in Colorado's Gunnison National Forest. Paul loved that he could find butterflies to watch right outside the door and field research sites in the nearby mountain
meadows. Anne learned to fly fish in the Colorado rivers. Their
daughter, Lisa, and sometimes her friends, accompanied Paul
and Anne to Gothic, where Lisa worked as a field assistant, fol-

lowing female butterflies, for instance, for twenty minutes at a time to see if they might lay an egg that could be collected. Over the years, Ehrlich would take daily morning walks through the mountain meadows with Harte and Holdren. Together the scientists would also climb thirteen-thousand-foot Colorado peaks and spend evenings drinking wine and telling stories. It was an extraordinary, if perhaps counterintuitive, setting from which to contemplate the collapse of society and the ruination of Earth.[32]

As Ehrlich and his friends discussed dire threats to the natural environment, they agreed that inviolable biological and physical limits meant that governments had to act aggressively to address the human population problem. Otherwise harsh laws of nature would bring society and the planet's ecosystems crashing down. To avoid this worst-case scenario, experts should determine the "stable optimum population size for the United States," Ehrlich wrote in *The Population Bomb,* and government should adopt policies to achieve that optimum number. His own view of a sustainable human population was radically small, expressed variously as approximately 17 percent (around 600 million) or, more commonly, 40 percent (1.5 billion) of the 1970 population of 3.7 billion people worldwide.[33]

Ehrlich was not alone in making this leap from biology to political advocacy. In an influential 1968 essay entitled "The Tragedy of the Commons," biologist Garrett Hardin warned that environmental problems could not be left to the "invisible hand" of Adam Smith. Hardin told a fable about a common grazing field in which all the farmers would keep adding their own cattle to the field for their individual benefit, regardless of impact on the degrading lands. Following the same logic, Hardin said, factories poured pollution into the air and water, reaping a private benefit while spreading the public harm. Every family also had an incentive to reproduce even if overpopulation reduced the general prosperity. Hardin argued that property rights needed to be redefined, from the right to pollute to the right to have as many children as one desired. The "freedom to breed is intolerable," Hardin said. "Coercion is a dirty word to most liberals," Hardin conceded. But to save itself, society needed to embrace "mutual coercion mutually agreed upon." Hardin's characterization of the "tragedy of the commons" became a pivotal trope in the environmental movement, used by Ehrlich and others to explain the destruction of unregulated common property resources, from fisheries to clean air.[34]

Ehrlich wanted to do more than simply talk about the need for action—he wanted to make things happen. Shortly after publishing *The Population Bomb*, Ehrlich joined with colleagues to launch an organization called Zero Population Growth to advocate for population control in the United States. The new organization reflected Ehrlich's belief that if the nation wished to promote population limits overseas, it had to get its own house in order. Plus, Americans consumed so much that each American born had a disproportionate impact on resource consumption. Ehrlich, the Yale biologist Charles Remington, and Richard Bowers, a Connecticut lawyer and conservationist, dreamed up Zero Population Growth after a game of squash in New Haven. Prominent biologists such as Garrett Hardin, Harvard's Edward O. Wilson, and George Woodwell from the Brookhaven National Laboratory joined the organization's board of directors. By April 1970, following Ehrlich's appearances on the *Tonight Show*, the organization had grown to nearly one hundred chapters across the United States. The concept of "ecology" had become a "politically potent word," Ehrlich wrote to members of the Ecological Society of America in a letter describing Zero Population Growth. The new organization would "make clear to Americans the intimate connection between runaway population growth, our 'cowboy' economy, and the deterioration of our environment." Scientists had a responsibility to correct all the "uninformed" talk.[35]

Wading into American politics inevitably meant that Zero Population Growth would have to tackle controversial issues such as birth control, abortion, and women's rights. Ehrlich's push for measures to reduce population growth thus drew on the 1960s sexual revolution and efforts to separate the pleasure of sex from reproduction. As a biologist, Ehrlich did not view sexual intercourse as anything sacred. He attacked "sexual repression" and celebrated sex as "mankind's major and most enduring recreation." Ehrlich waged an aggressive campaign against Pope Paul VI's 1968 encyclical "Humanae Vitae," which affirmed the Catholic Church's traditional proscription of most forms of birth control. Zero Population Growth continued this fight in the early 1970s, advocating forcefully for abortion rights and access to contraception. In California, Zero Population Growth sought to help pass a proabortion ballot initiative. Ehrlich, who served as the organization's first president, urged the legalization of abortion and removal of restrictions on contraception in the interest of population control. Ehrlich mocked the association of a fetus with a human being as "confusing a set of blueprints with a building." After New York passed a liberal abortion law in 1970, Charles Wurster, a founder of the nonprofit Environmental Defense Fund, wrote exultantly to Ehrlich, "I wouldn't have dreamed this could have happened so fast! This bill is now LAW in the State of New York." Wurster considered population growth the "grand-daddy problem of them all" and, like many environmentalists, saw ready access to abortion as an important tool for preventing unwanted births. Ehrlich also urged men to take responsibility for birth

Paul and Anne Ehrlich meeting with an editor to discuss their 1970 book,

Population, Resources, Environment. Courtesy of Paul Fusco, photographer,

LOOK Magazine Collection, Library of Congress. Prints and Photographs Division.

control, including through vasectomies. Ehrlich publicized his own vasectomy and even included the fact that he had been sterilized in his byline for a 1970 article in *Audubon* magazine. He also served on the board of directors of the Association for Voluntary Sterilization, which advocated the surgical procedure. Other population control advocates followed Ehrlich. Fred Abraham, a member of the UCLA psychiatry department and president of Zero Population Growth in Los Angeles, described his vasectomy surgery in the *Los Angeles Times* in 1970. *Time* magazine noted the public campaign but dismissed it as a fringe movement, saying, "It is obvious that few Americans will imitate Paul R. Ehrlich and some of his young disciples."[36]

Ehrlich and Zero Population Growth also took positions on the role of women in the workplace and the number of children that families should have. Women's rights were not a major focus for Anne and Paul, and in their own marriage, they filled fairly traditional roles. Paul expected Anne to stay at home to care for Lisa, make dinner, pack his clothes for trips, and otherwise care for the household. Anne's drawing, writing, and, later, her public speaking, made her much more than a home-maker, but her professional life developed as an offshoot of, and was largely subordinated to, Paul's work. Anne considered herself a feminist, but it was not her main interest. Where women's rights were connected to population issues, however, the Ehrlichs supported them aggressively. Encouraging women to work, they thought, would lower family size, so they called for free or low-cost childcare as way to get women out of the house and into the workplace. Anne Ehrlich wrote to a col-league in 1969, "An amazing number of women have another baby because they have nothing else to do." Roger Revelle, director of the Harvard Center for the Study of Population, agreed that women needed more career opportunities as alter-natives to being mothers. "There is one movement on which I am a radical—the Women's Liberation movement," Revelle declared. At the same time, the primary focus of Zero Popula-tion Growth and the Ehrlichs on population sometimes meant that they took positions that risked making life more difficult for women. Their call for federal policies that would discourage larger families, including "luxury taxes" on diapers, baby bot-tles, and baby foods, could seem harsh. The Ehrlichs also called for tax reform to discourage reproduction, including a change from tax deductions to tax increases for children, and proposed lotteries and awards for childless couples. Wes Jackson, a plant biologist and founder of the Land Institute in Kansas, agreed that couples should pay "the least tax if they have no children" and "through the nose" if they have three children. Jackson also called for taxes on baby bottles, blankets, cribs, and other par-enting supplies.[37]

While explicitly demanding higher taxes on larger families, the biologists skirted the edge of calling for direct government

prevention of "extra" births. Ehrlich, Hardin, and others refused to rule out coercion as a potentially necessary and justifi-
able option, but they recognized that it was not politically feasible. Ehrlich rejected the idea that parents had an "inalienable
right" to reproduce and said that perhaps an across-the-world
limit of two children per family would be the most equitable
approach. Ehrlich floated hypothetical ideas about mandatory
sterilizations or temporary infertility imposed through pills or
public drinking water. Yet he generally shied away from calls
for direct government controls. Pointing to the furious and sus-
picious response to fluoride supplements in public water supplies, Ehrlich declared that "society would probably dissolve
before sterilants were added to the water supply by the government." Even though Garrett Hardin had written that "the freedom to breed is intolerable" and believed that voluntary birth
control was doomed to failure, Hardin warned against coercive
methods for strategic reasons in 1970. Hardin thought effective coercive techniques still needed to be invented. In the
meantime, Hardin said, "we ought not say anything foolish
about coercion." Instead, the population control movement
should focus on discouraging births and expanding access to
abortion and birth control. At one point, Ehrlich sought funding for a small-family campaign by Zero Population Growth
modeled on well-known antismoking advertisements.[38]

Critics could not quite be sure where Ehrlich, Hardin, and
others stood on the idea of coercion. In *The Population Bomb*,
Ehrlich called for the creation of a "powerful governmental
agency" to coordinate "whatever steps are necessary to establish a reasonable population size in the United States." This
"Department of Population and Environment" would, among
other activities, fund research to develop new birth control
techniques, possibly including "mass sterilizing agents." "Not

all compulsory methods are necessarily horrifying," Ehrlich declared in a 1969 speech, in which he also said, "It is clear that
governments, including eventually a world body, must under-
take the task of regulating the population size just as they now
attempt to regulate economies." These ideas provoked resistance and leaders of Zero Population Growth struggled to dis-
associate the organization from calls for "compulsory birth
control." Executive Director Shirley Radl assured Ehrlich in
April 1970 that the organization was "acutely aware of what is
damaging to our issue and what is beneficial." She knew that
"to discuss hard-line compulsory birth control is totally taboo."
Radl insisted that "parenthood should be a privilege and not a
right." But she acknowledged that advocating coercion would
be "detrimental." The new generation of youth did "not take
kindly to force. I know that the climate in California is just not
right for any talk of compulsion." Ehrlich and Zero Population
Growth instead promoted popular education, voluntary birth
control, and more moderate policy reforms. Yet Ehrlich was a
scientist and a debater, drawn to subtle distinctions of logic and
abstract points of principle. He stubbornly defended the idea
that dire circumstances could justify drastic coercive action. If
the world took no action, he warned, "it will wake up one day
to find out that *compulsory* birth control offers the only hope of
survival." Ehrlich and other population control advocates thus
left themselves open to criticism from both the political left,
concerned with the impact of coercive population measures on
the poor and minorities, and the right, fearful of oppressive
government meddling in family affairs.[39]

Paul Ehrlich always saw population issues as part of a
larger environmental crisis, and in *The Population Bomb* he devoted considerable attention to the problems of toxic pesticides

and industrial pollution. After the book's runaway success, Ehrlich sought to use his fame and his credentials as a scientific expert for broader political influence. Ehrlich emphasized issues more than party affiliation. His public endorsements largely favored Democrats, but he also worked closely with liberal Republicans who shared his environmental concerns, such as Peter McCloskey. McCloskey represented the Ehrlichs' congressional district in California, and Paul and Anne strategically registered as Republicans so that they could vote in Republican primaries in support of him. In 1968, Ehrlich and eleven other ecologists backed Vice President Hubert Humphrey's campaign for president. In a public letter to three thousand scientists, the ecologists blamed "the crisis of food, water, air, space, of human numbers, dignity and of survival itself" for causing "political instability, disaffection of youth, and the decline of human values." Ehrlich and his coauthors acknowledged that some ecologists might wish to avoid politics. But they rejected that attitude as irresponsible. "Can the nation afford this position, or must every ecologist search his conscience, examine the record and his political beliefs, and work to elect the candidates who offer the best hope of guiding us in this monumental crisis?" Ecologists, Ehrlich and his coauthors insisted, could contribute by writing letters to the editor, assisting candidates, and educating policy-makers. They needed to force discussion of the "most fundamental issues." If their efforts were effective, then "after the election ecologists will be in a stronger position to advise the new administration."[40]

After Republican Richard Nixon decisively defeated Humphrey in the 1968 election, Ehrlich initially wrote to the president-elect to offer to work with his administration. "Biologists throughout the world are uniformly concerned" about the lack of governmental attention to the "population-food-environment crisis," Ehrlich said. He wanted to help Nixon get the "dramatic action which is necessary if we are to survive." Ehrlich's open attitude toward the new administration did not last long. When Nixon appointed Alaska governor Walter Hickel to be secretary of the interior, Ehrlich urged scrutiny of Hickel's prodevelopment record in Alaska. In a press release issued by the Stanford University News Service, Ehrlich denounced Hickel, saying that he "neither understands nor appreciates the basic principles of conservation or the need to preserve man's environment from destruction by his own carelessness." Hickel's appointment, Ehrlich wrote privately to a friend, was "enough to make any conservationist's toes curl."[41]

Ehrlich also sought to influence state politics and to advance specific policy proposals through the California legislature. In 1969, Ehrlich sent a long list of legislative ideas about population and environmental concerns to California assemblyman John Vasconcellos, a Democrat who represented San Jose. Among Ehrlich's population ideas: discouragement of immigration to California; birth control for all, including the poor; more liberal abortion laws; and changes to tax laws that discriminated against single people and childless and small families. He also called for greater economic opportunities for women and low-cost daycare for children, both of which would help bring women into the workforce and lead them to have fewer children. On the environmental front, Ehrlich urged that California ban DDT and make it harder to buy pesticides and weedkillers. He called for further measures against smog, including the encouragement of public transport. "It is too bad the measure banning internal combustion automobiles . . . failed to pass," Ehrlich wrote of a bill actually proposed by a California state senator from Oakland. (The bill passed the state senate in 1969 but was defeated in the assembly.) Ehrlich suggested instead that California ban "any automobile with, say, more than 100 or 125 horsepower, with perhaps strictly supervised exceptions for people with a demonstrable need for more." Ehrlich also urged government planning to slow development in California and discourage suburban sprawl. Frank Olrich, Vasconcello's assistant, responded that some of Ehrlich's proposals, such as the attack on the internal combustion engine, were "far out—in terms of California politics, and the level of awareness here in Sacramento. However, a few of them look good in all respects." Ehrlich replied that he "knew some of them were too far out for immediate enaction [sic], but . . . I felt they were worth including. They may not seem so far out two or three years from now." Ehrlich hoped to set a course for the country's redirection.[42]

Ehrlich was right that the nation's environmental politics were changing rapidly. A dramatic oil spill off the coast of Santa Barbara, which occurred just weeks after Nixon's inauguration in 1969, gave new urgency to environmental problems. Televised images of oil-soaked birds dying on the California beaches galvanized opposition to industrial pollution. Anti-oil activists in Santa Barbara quickly organized a protest group called Get Oil Out, or GOO, to oppose oil drilling along the California coast. Local organizers issued the Santa Barbara Declaration of Environmental Rights, calling for "a revolution in conduct toward an environment which is rising in revolt against us." In addition to the threat of oil pollution, the declaration deplored litter, air pollution, species extinction, and lost open space, and called for a new ethics to "govern man's contact with all life forms." Six months after the Santa Barbara spill, the nascent environmental movement gained further impetus when Cleveland's Cuyahoga River, awash in refinery waste and other debris, caught fire. The blaze was not the first, nor the worst, fire on the Cuyahoga, but it exacerbated fears that Americans were laying waste to the natural environment. Predictions that Lake Erie was "dying" due to oxygen depletion and algal blooms caused by pollution similarly fed fears of environmental degradation.[43]

President Nixon responded by shoring up his environmental credentials to outflank potential 1972 Democratic rivals like Henry Jackson and Edmund Muskie. In July 1969, just a year after Ehrlich published *The Population Bomb*, Nixon gave a speech calling global population growth a "world problem which no country can ignore." He was the first American president to give population such prominent attention, a sign that Ehrlich's public campaign was bearing fruit. Nixon warned that growth in human numbers threatened to outstrip economic development. He vowed support for voluntary population and family planning efforts in other countries. Nixon also declared that population growth posed "serious challenges" for the United States. He pledged greater federal funding for family planning programs and for study of the relationship between population growth and environmental quality. Nixon also called for the creation of a commission to examine future population trends and their consequences in the United States. The Commission on Population Growth and the American Future, which started its work in 1970, would be headed by John D. Rockefeller III, a leading financial backer of Planned Parenthood, the Population Council, and other population organizations, as well as the grandson of the founder of Standard Oil.[44]

Nixon continued his aggressive campaign on environmental issues into 1970, determined not to cede any political ground to the Democrats. Congress and the president together embarked on an unprecedented burst of bipartisan policymaking, born as much from political rivalry as common cause.

The new wave of federal legislation began with the National Environmental Policy Act, a broad statement of environmental values that called for "productive and enjoyable harmony between man and his environment." The law required federal agencies to assess the environmental impact of their actions. It also created a new Council on Environmental Quality to advise the president and help coordinate federal policies. Nixon initially preferred a less vigorous environmental advisory council that he already had created by executive order, but he had to support the law, or face heavy criticism. In signing the National Environmental Policy Act on January 1, 1970, his first official act of a new decade, Nixon declared himself convinced that the "1970's absolutely must be the years when America pays its debt to the past by reclaiming the purity of its air, its waters, and our living environment. It is literally now or never."[45]

Moving aggressively to seize control of the issue, Nixon declared in his January 1970 State of the Union address that environmental restoration was a "cause beyond party and beyond factions." The "great question of the seventies" was whether Americans would "surrender" to environmental deterioration or make "our peace with nature." The nation could not let urban Americans be "choked by traffic, suffocated by smog, poisoned by water, deafened by noise, and terrorized by crime." The United States should pursue balanced economic growth that preserved "the quality of life" for Americans. Nixon released a special environmental message to Congress in early February 1970 announcing a sweeping series of legislative proposals and executive orders to address air and water pollution, solid waste management, and parks and recreation. Cleaning up the environment—going beyond mere conservation to environmental "restoration"—called for a "total mobilization by all of us." Nixon sought to frame environmental problems as a collective challenge facing all Americans. He rejected a "search for villains" and "evil men," and instead pointed to common carelessness and environmental neglect. Nixon's position fit the growing focus on personal responsibility and an unsustainable American way of life. As Walt Kelly's cartoon character Pogo famously declared of the growing pollution problem later that year, "We have met the enemy, and he is us."[46]

Nixon and his Washington rivals were just keeping up with the surging popular interest in environment protection that Paul Ehrlich had helped to create. In April 1970, more than twenty million Americans took to the streets to call for environmental action in the first Earth Day. Paul Ehrlich served on the small steering committee of only eight people for the national Earth Day organization. Wisconsin Democrat Gaylord Nelson, a leading environmental champion in the Senate, had proposed the idea of a national environmental teach-in, and California Republican Peter McCloskey, an environmental leader in the House of Representatives and Ehrlich's congressman, embraced the idea. Ehrlich's former Stanford student, Denis Hayes, led the national organization that struggled to coordinate many independent local Earth Day activities. At a large Northwestern University "teach-out" leading up to the official Earth Day, Ehrlich spoke to a crowd of more than eight thousand. The Northwestern event showcased nine speakers for more than four hours and then broke up into nineteen discussion groups that lasted until dawn. One highlight of the event occurred when twenty-five American Indians, including two in full regalia, interrupted the proceedings by seizing the podium and demanding that the university stop aiding companies that have been "polluting us to death." At a talk in Southern California, Ehrlich said the people of "overdeveloped countries" were "the looters and polluters of the planet. . . . We have to change

Paul Ehrlich, speaking to thousands at Iowa State University, 1970. Courtesy of Special Collections Department/Iowa State University.

our way of life or we're going to die." The vigorous critique of American practices by Ehrlich and other commentators had become a new mainstream perspective. The *New York Times* declared the April 22 Earth Day as American an event as Mother's Day. "Conservatives were for it. Liberals were for it. Democrats,

Republicans and Independents were for it . . . no man in public office could be against it."[47]

The political momentum behind pollution control overwhelmed Washington's traditional resistance to change. Three months after Earth Day, Nixon pushed through a restructuring of the federal government's anti-pollution programs to create a new, unified Environmental Protection Agency (EPA). Congress then dramatically expanded the new EPA's mandate with federal laws requiring sharp improvements in air and water quality. At the same time, the federal courts strictly interpreted new requirements for environmental impact statements, and empowered independent environmental law organizations to intervene in controversial development projects. Litigation over potential environmental impacts now delayed huge development projects, such as the Alaskan oil pipeline, for years. Other projects, such as the South Florida jetport that John Harte had opposed, were canceled altogether. The new laws and the court rulings profoundly altered the federal government's relationship to the environment and, therefore, the economy. It was a stunning and quick victory for the environmental movement, and the most profound and sudden shift in the relations between businesses and government since the end of World War II.[48]

Paul Ehrlich, however, was not satisfied, and neither were many environmentalists. Although President Nixon described the global population explosion as a "rush toward a Malthusian nightmare," Ehrlich called for even more aggressive action to counter population growth and environmental degradation. In January 1970, Ehrlich described as "hilarious" Nixon's pledge of ten billion dollars for water pollution control. Ehrlich estimated instead that the nation needed to spend fifty to sixty billion dollars *per year* to bring pollution under control. Despite

Nixon's signing of the National Environmental Policy Act and his reorganization of the Environmental Protection Agency, Ehrlich criticized Nixon in the lead up to the 1972 election for his "total lack of action on truly critical environmental problems, and his hopeless positions on population policy." Ehrlich, who opposed the Vietnam War, also considered Nixon "one of the first major eco-criminals" for what Ehrlich considered an "ecocidal" attack on the people and environment of Vietnam. Other environmentalists shared Ehrlich's disappointment and hostility toward Nixon, a feeling encapsulated in a 1972 book entitled *Nixon and the Environment: The Politics of Devastation.* Despite the many changes underway, environmentalists felt that Nixon stood in the way of more sweeping reform, offering more rhetoric than action.[49]

Nixon, in turn, grew frustrated by the lack of political advantage that he found in the environmental arena. He thought that environmentalists like Ehrlich made extreme demands that he could never hope to satisfy. After the 1970 midterm elections, Nixon began to shift away from environmental policy. "The environment is not a good political issue," Nixon told his chief of staff, H. R. Haldeman. "I have an uneasy feeling that perhaps we are doing too much." Nixon increasingly focused on the high economic costs of environmental policy. He worried that the new environmental laws impeded economic growth, and he publicly warned against "ecological perfection at the cost of bankruptcy." Privately, Nixon complained, "Some people want to go back in time when men lived primitively." He told a group of auto executives, again privately, that environmentalists were "enemies of the system." Nixon rejected the concerns of his own population commission in early 1972; as part of his strategy to win over Catholics in advance of the 1972 election, Nixon instead emphasized the threat of unrestricted

Richard Nixon, San Clemente, California, January 1971.

Courtesy Richard Nixon Library.

access to abortion. Publicly, Nixon continued his efforts to neutralize environmental issues as a political concern, but he increasingly focused his attention on international diplomacy, including a landmark visit to the People's Republic of China.[50]

As the 1972 election approached, Ehrlich joined the National Committee of Environmentalists for McGovern. He backed the Democratic nominee somewhat grudgingly, however, refusing to endorse McGovern during the primaries because his stance on population was "not strong enough." Ehrlich particularly pressed McGovern to embrace liberal positions on abortion and birth control. Recognizing the political risk of linking abortion with population control, Ehrlich sought to distinguish between the two. The "problem of population control is primarily one of changing people's values so that they *want* fewer children," he explained. The availability of contraception and abortion simply made it "easy and safe for couples to have the smaller families they have decided on." Given sufficient assurances, Ehrlich finally endorsed McGovern publicly in the fall of 1972. He urged McGovern to attack Nixon on environmental issues, and complained that he was not called upon to help more in the campaign. "I strongly suspect that my association with you would be a political plus with selected audiences in California." Ehrlich thought that he knew where public sentiment lay and that environmental problems would have broad appeal as a campaign issue. Ehrlich drafted his own press release in October criticizing Nixon for ignoring environmental injustices inflicted on blue-collar workers, the poor, and racial minorities. "They must work and live in the smog and filth, they must labor in farm fields exposed to high risks of pesticide poisoning—they can't live upwind of the pollution along with Nixon's millionaire cronies," Ehrlich said. "No group stands to gain more from ecological sanity than blue collar workers." Despite such attacks from Ehrlich and other environmentalists, Nixon had successfully protected his environmental flank through his many accomplishments between 1969 and 1972. Environmental issues were neutralized in the campaign, which focused instead on the Vietnam War, cultural issues, and the electroshock therapy of McGovern's first vice presidential running mate. Nixon trounced McGovern in the 1972 election and Ehrlich found himself once again on the outs with the presidential administration.[51]

As Ehrlich's public stature grew, he and his family paid a personal price. After the wave of publicity from his turn on *The Tonight Show*, the family received death threats as well as attention from mentally unstable people. Anne told Lisa to keep the curtains closed so that outsiders could not see in the house to shoot at them. They purchased an early alarm system for their house, and they tried not to let people know where they lived. It was a tumultuous time in Northern California. The Stanford campus and Palo Alto area was in turmoil over the Vietnam War. One summer, a mentally ill woman who had grown obsessed with Ehrlich broke into their house and began living there with her dog, while the family was away at the Colorado field station. When the police came to investigate, they found the Ehrlich's house in disarray, with piles of papers and books in apparent chaos. The place had been ransacked, they thought. It turned out, however, in what would become a family joke, that this was just the way Paul and Anne lived.[52]

Few scientists had the internal constitution or rhetorical skills to play the public role that Ehrlich did. Some of Ehrlich's scientific colleagues described feeling "schizophrenic," torn between professional responsibilities and their personal reticence, on the one hand, and the "moral bind" that overpopulation placed on them, on the other. Others questioned whether Ehrlich's provocative style served him well, and criticized his apocalyptic rhetoric. As Eugene Odum, a leading ecologist, wrote to Ehrlich in 1970, "while some of us like yourself must remain 'highly visible,' we have also got to encourage many other ecologists to back up this visibility with what we might call real credibility." In a tough review of Paul and Anne Ehrlich's 1970 *Population, Resources, Environment: Issues in Human Ecology*, Roger Revelle, a leading oceanographer and the director of Harvard's Center for Population Studies, called Ehrlich the "New High Priest of Ecocatastrophe." The "emotional and quasi-religious force" of Ehrlich's writing, Revelle wrote, was not likely to "lead to the hard thinking and effective action which the overwhelming issues so urgently demand." Revelle particularly complained about the Ehrlichs' rampant use of "apocalyptic adverbs and adjectives": "staggering, sobering, disaster (three times), enormously, drastically, catastrophic, dramatically, tremendous, highly lethal, extremely dangerous (twice), especially virulent, more severe, extremely fortunate, extremely vulnerable, almost total, high potential, renewed spectre, not gruesome enough, colossal hazard, biological doomsday, superlethal, disastrously effective." And these appeared in only *four* of the book's three hundred–plus pages. In a more friendly manner, Ehrlich's mentor Charles Birch jokingly called Ehrlich "Dear Billy," saying that "just for a moment I thought I was writing to Billy Graham," the evangelical preacher. Birch continued, "Your last letter read just like one of Billy's sermons with the impending Armageddon approaching and the plea to prepare now while we have time." The theological association was not too far-fetched. In late 1968, just after *The Population Bomb* was published, Ehrlich delivered Sunday sermons in San Francisco's Grace Cathedral and at Stanford's Memorial Church.[53]

Ehrlich's sweeping rhetoric and sharp tongue thus motivated many but also alienated others who thought his predictions of doom and gloom were vastly overstated. Even his sympathizers expressed concern. John Lear, an editor at *Saturday Review,* for whom Ehrlich and Holdren wrote a regular column in 1970 and 1971, reported receiving telephone calls from leading scientists, as well as former Ehrlich students, to criticize Ehrlich's essays. Lear rewrote part of one 1971 Ehrlich and Holdren column in order to scale back their claims, telling

them, "We have had so many complaints about oversimplification of the problems we discuss that I consider some protection
necessary from time to time." Ehrlich's acid tongue attracted
attention and stirred controversy, as when Ehrlich called Jack
Williams, the governor of Arizona, a "clown" and "moron" for
his views on environmental and population issues. Moral conviction and self-confidence underlay Ehrlich's willingness to
call his opponents "idiots" and fools. As he wrote in a 1970 letter to William Draper, head of the Population Crisis Committee, "I am busily writing outrageous things in a continuing
attempt to change the unchangeable." Some of his fellow scientific advocates shared Ehrlich's feeling of righteousness
mixed with desperation, writing letters to their "colleagues in
arms," signed "In the name of the planet." But most others
did not. "The thing I would trust Ehrlich with is butterflies,"
commented Philip Hauser, a demographer at the University
of Chicago.[54]

While some critics attacked Ehrlich for oversimplification
and excessive pessimism—and disliked his harsh tone—others
from the political left criticized his obsession with population
growth as the root cause of environmental problems. Barry
Commoner, a plant physiologist at the University of Washington in St. Louis, cared little about population issues. Commoner thought that poverty, technology and science deserved
more attention. Commoner had helped publicize the dangers
of above ground nuclear testing in the 1950s, warning that
strontium-90, the radioactive isotope, was entering the food
chain, and people's bodies. In his 1971 best seller, *The Closing
Circle,* Commoner argued that neither population nor affluence caused the growing assault on the natural environment.
Rather, new technologies and scientific creations, including
plastics and detergents, broke with natural biological cycles

and created dangerous new types of pollution. Corporate pursuit of profits drove the worsening state of the environment, Commoner argued, not the growth in human population. Calls for population control were like "attempting to save a leaking ship by lightening the load and forcing passengers overboard." Commoner argued instead that there was something "radically wrong" with capitalism and its embrace of new technologies. Commoner made his own simplifications, of course, in attacking Ehrlich's emphasis on overpopulation. As Ehrlich pointed out in his retort to Commoner's "dreadful book," Commoner's focus on chemicals and technology minimized other important human impacts on the environment, such as the clearance of land for agricultural settlement, that predated even the industrial revolution.[55]

The fight between Commoner and Ehrlich to define the fundamental causes of environmental problems grew increasingly personal as the two proud scientists lambasted each other in print and in public forums. Commoner infuriated Ehrlich and Holdren when he preemptively published Ehrlich and Holdren's critique of his book, without their authorization, in the magazine *Environment*, along with Commoner's own rebuttal. (Ehrlich and Holdren's piece had been scheduled to appear in the *Bulletin of the Atomic Scientists* at a later date, but had circulated fairly widely before publication.) Their conflict reached a low point at the United Nations environment conference in Stockholm in 1972. At the nongovernmental forum that occurred alongside the official conference, Ehrlich participated in a panel discussion with two Commoner sympathizers. As the discussion got under way, five additional anti-Ehrlich activists representing third world concerns came onto the stage to criticize Ehrlich, turning the session into what seemed like an ambush. Commoner sat in the audience in a balcony overlooking the proceedings, from time to time passing down handwritten notes and questions to his allies. Ehrlich finally called out, "Come on out, Barry baby!" to try to goad him into a direct argument, but Commoner declined.[56]

Ehrlich's clash with Commoner highlighted the competition between leading spokespeople and also illustrated how Ehrlich's relentless focus on population left him open to criticism from the left. Advocates for the poor and people of color joined Commoner in slamming Ehrlich's calls for population control. Some thought that the population movement was motivated by a eugenic quest for racial improvement. "The black power advocates won't buy us at all," wrote one Zero Population Growth member in New Jersey. "They are convinced we are interested only in genocide." Their fears were not unreasonable. John Holdren's inspiration, Harrison Brown, had worried about the "genetic soundness" of the human species. Frederick Osborn, a cousin of the Fairfield Osborn who wrote *Our Plundered Planet,* helped found both the American Eugenics Society in 1926 and the Population Council in 1952. In dozens of states, eugenic policies had resulted in the sterilization of thousands of largely poor and minority women in state mental institutions. At a national meeting on "Optimum Population and Environment" in 1970, forty African American participants walked out in protest. The protestors refused to "legitimize a preconceived vicious plan of extermination" aimed at reducing the population of blacks, other non-whites, the poor, and immigrants. "We cannot participate in our own destruction," declared Dr. Alyce Gullattee, a staff psychiatrist at St. Elizabeth's hospital in Washington, DC. Gullattee criticized several ideas expressed by conference participants: that having children is a privilege and not a right; that children might be given antifertility vaccines and then given an antidote later to allow them

to procreate; and that people who lived in ghettos were responsible for their desperate condition. Gullattee argued that birth
control efforts would not affect affluent white families. Instead,
she said, new laws would "easily control a poor black family,
especially those dependent on government assistance." Leroy
Richie, who directed the black student program for the National Urban League, warned more starkly, "They are coming to
get us and we've got to stop them." The conference protest also
reflected broader African-American concern that the increasing popularity of the environmental movement threatened efforts to address entrenched social inequality and racism. Richard
Hatcher, the mayor of Gary, Indiana, criticized environmentalists for "distract[ing] the nation from the human problems of
black and brown Americans." The "war on pollution," the
Urban League's Whitney Young declared, "should be waged
after the war on poverty is won."[57]

Paul Ehrlich thought of himself as a social justice advocate
concerned with the poor and minorities. Ehrlich himself did
not believe in the idea of racial differences. He thought that the
biological category of race provided little useful information to
distinguish among groups within society. In a 1967 review of a
synthesis of modern biology, Ehrlich wrote that "the application of the term 'race' to man is very dangerous, and it is probably unwarranted for other animals and plants." While Ehrlich
did not take a particularly active role in the civil rights movement in the 1960s, he did play a small part in trying to desegre-
gate restaurants in Lawrence, Kansas, while he was in graduate
school in the 1950s. A black Jamaican scientist came to visit the
laboratory one time but was refused service at the hotel restaurant, instead subsisting on vending machine candy and other
food over the weekend. Ehrlich and some of his scientific colleagues organized a protest of his treatment. More generally,

Ehrlich criticized America's "racist society" for the inequality of educational opportunities available to children divided by skin color. And he spoke forcefully about the problems of the ghetto and social inequality, albeit often in an environmental or population framework. He believed that environmental problems were particularly insidious threats to disempowered groups, and he sought to show that efforts to halt population growth were consistent with social equity.[58]

As the closing speaker for the 1970 population and environment conference, Ehrlich forcefully defended the event, and the population control movement more generally, against charges of being anti-poor and racist. But he also praised the protestors for drawing attention to urgent problems, such as inadequate education, social inequality, and urban decline. Ehrlich said that the slums should be the first priority for environmental improvement. The following month, in an essay published in the National New Democratic Coalition Newsletter in July 1970, Paul and Anne Ehrlich explicitly addressed the relationship between population control and "genocide." The Ehrlichs conceded that the perception among people of color that population control policies were genocidal was not "entirely unjustified," given the attitudes of some advocates, "who seem mainly interested in controlling other people's populations." The Ehrlichs called for government population control programs that would first reduce the birthrates among "affluent white Americans." Population growth among the affluent in the United States, Europe and other countries posed the greatest threat. "In terms of degradation of the environment, the birth of each American child is 50 times the disaster for the world as the birth of a child in India," they wrote. "Similarly, poor people in the U.S. have far less power to loot and pollute than does the average American." The Ehrlichs urged attention first to population growth in the United States among affluent and middle-class whites, the greatest force for environmental destruction.[59]

Despite Ehrlich's blunt critique of middle-class American consumers, the question of racial bias in the population movement continued to plague him. The population movement's focus on reducing family size raised the specter of social control of supposedly undesirable or politically vulnerable groups of people. Ehrlich dismissed the idea of race and criticized American racism, yet he struggled to defend himself and his ideas. In 1974, Students for a Democratic Society and other groups formed a "Committee Against Racism" and criticized *The Population Bomb,* along with writings by other scholars. Ehrlich would spend a significant part of the 1970s seeking to persuade the public that population control, and its newer political manifestation, limitations on immigration, rested on a solid scientific and rational foundation.[60]

Ehrlich's political vulnerabilities illustrated the complicated relation between the emerging environmental movement and American liberalism, with its emphasis on economic growth spurred by government action and on rights-based social equality. Questioning the limits of population, technology, and economic growth ran counter to the big-government liberalism of people like California governor Pat Brown, who oversaw the construction of massive freeway and water projects in the 1960s, as well as the expansion of the state university system. Ehrlich's critique of excessive consumption similarly contradicted liberal efforts to drive prices down to facilitate mass consumption. Ehrlich's suggestion that childbearing rights might be limited or controlled by the state greatly concerned civil rights advocates as well as conservatives. Environmentalism thus challenged the ideas of the American political left as well as the right. Environmentalism had the potential to threaten privileged groups, through economic regulation and restrictions, and also safeguard privilege, by constraining developments that threatened private lands and interests. It was partly this complex political positioning in American culture, as well as fractious competition among individuals such as Ehrlich and Commoner, that kept the environmental movement from raising any one unifying leader to represent environmentalism. Environmentalism has never had a Martin Luther King Jr. figure to set and hold a moral compass for the movement and the nation. Ehrlich's deep continuing commitment to his scientific research and teaching provided him with a rhetorical platform as a public intellectual but also limited his role. He served on advisory boards but avoided official management or leadership positions. He was a thought leader, not a lawyer or a political organizer, in a movement increasingly defined by law and by the fierce struggle over the new rules being established during the first years of the decade. Ehrlich's focus on large-scale change, and on broad threats to humanity, meant that he stood outside increasingly technical and often compromising negotiations in Washington and state capitols. Ehrlich, along with peers such as Commoner, instead synthesized intellectual frameworks for thinking about the ecological crisis and urgently called Americans to action.

CHAPTER TWO   **Dreams and Fears of Growth**

W

hile Paul Ehrlich spent the winter of 1970 racing around the country speaking on national television and radio and to mostly adoring crowds, Julian Simon stayed at home in Urbana, Illinois. No one cared much about what the little-known professor of economics and marketing thought of environmental problems. In late February, however, the prominent psychiatrist Robert J. Lifton canceled a planned speech on the youth movement to the YMCA/YWCA Faculty Forum in Urbana. Julian Simon agreed to speak in Lifton's place. Simon's talk, "Science Does Not Show There Is Over-Population," laid out the themes he would pursue over the next few decades. "I view the population explosion not as a disaster, but as a triumph for mankind," Simon boldly declared. "Whether population growth is too fast or too slow is a value judgment, not a scientific one."[1]

Despite the small forum, Simon's contrarian perspective drew local attention in the student newspaper and among his faculty colleagues. Two months later, during the "Environmental Crisis Week" teach-ins that surrounded Earth Day in April,

Simon was invited to provide a contradictory viewpoint on a Tuesday evening panel about "The Population Program." This time he had a sizable audience. In a crowded auditorium at the University of Illinois, Alan Guttmacher, president of Planned Parenthood, gave a keynote lecture on family planning and the impact of population growth on the environment. Simon followed Guttmacher and expressed his skepticism that population growth and resource scarcity posed genuine problems. University of Illinois biology professor Paul Silverman, one of Simon's colleagues, then rose to speak. With Simon sitting on the podium beside him, Silverman ranted for twenty minutes about Simon's remarks, ignoring Guttmacher. Silverman called Simon a "false prophet" whose optimism about the world food supply "lacks scholarship or substance." He portrayed himself and Guttmacher, as well as Paul Ehrlich, by contrast, as fighting "for the future of the world." Simon fumed over the attack, complaining that the organizers had promised that each speaker would address the topic of the day, not attack each other. He also bore a grudge: two weeks later, at a faculty party, Simon accosted Silverman and threw a drink in his face. Scuffling ensued. It was an awkward, even violent, start to Simon's public life.[2]

Simon's moment in the hot seat on the podium at his own university reflected his position in the 1970s as a skeptic challenging Ehrlich's apocalyptic prophesies. Simon's doubts were newfound: he, too, initially had assumed that rising populations posed a dire economic threat. In academic articles in the late 1960s, Simon sought to justify investment in family planning on economic grounds and to suggest ways to improve marketing and information campaigns. Simon's early enthusiasm for population control, however, soon gave way to skepticism about whether population growth posed a real economic problem. Given the ambiguous economic impact of growing populations, Simon increasingly doubted whether economists and demographers could easily judge the relative morality of population growth.

During the early 1970s, Julian Simon's writing about the economic impact of population growth fit squarely within a framework other economists were building for understanding the factors that contribute to, and constrain, economic growth. Even as biologists like Paul Ehrlich highlighted intractable natural and physical limits to human activity, mainstream economists questioned the distinctive importance of natural resources. They instead emphasized human capital, technology, and innovation. Nature was just another factor in economic systems, they argued. Markets would successfully manage the depletion of resources by developing substitutes, moderating demand, and stimulating production. Julian Simon shared these views and developed them into a provocative critique of environmentalism and a sharp attack on Paul Ehrlich's apocalyptic view of the future.

"I first learned to say 'Do you want to bet?' as part of an argument about facts when talking to my father," Julian Simon wrote in his autobiography, *A Life Against the Grain*. "He would say outrageously wrong things in an authoritative fashion and refuse to hear any questions. There really was nothing I could say except 'Do you want to bet?'" Simon liked facts and data, and he enjoyed testing his theories against others on an empirical basis. He had a combative personality and relished the sharp give-and-take of rhetorical conflict. His academic work thoroughly absorbed him. Simon wrote or edited more than twenty books, more than a hundred academic articles, and an even greater number of opinion essays and general interest articles.[3]

In his polemical style, as well as his dedication to scholarship and to rational argument, Julian Simon greatly resembled Paul Ehrlich. This was no coincidence, since they came from similar Jewish communities in New Jersey. Simon's grandparents kept a hardware store in downtown Newark in the 1930s; his aunts lived above the store in a second-floor apartment. His own parents moved out to the Newark neighborhood of Weequahic, later made famous as the place where the neurotic Jewish protagonist of Philip Roth's *Portnoy's Complaint* spent his childhood. Weequahic, located near where the Newark airport is today, was largely Jewish before the war. Two-story clapboard houses lined the street, and a kosher butcher was on the corner. A milkman still delivered milk with a horse and wagon when Simon was a child. In 1941, Simon's parents moved with Julian, age nine and their only child, to Millburn, New Jersey, just a few miles down the road from Paul Ehrlich's home in Maplewood. Like the Ehrlichs, the Simons were part of a wave of upwardly mobile Jewish families dispersing into the suburbs surrounding Newark.[4]

Simon felt increasingly isolated after moving to Millburn, and he was keenly aware of his family's financial insecurity. He had left behind the bustle of immigrant life in Weequahic for the assimilation of the suburbs. He sometimes found himself the only Jewish kid around. His father struggled to find and keep a job after the move to Millburn, facing unemployment for long periods. Family members maintained a careful front to disguise their economic insecurity. For a time, Simon's father worked in New York City as a coffee roaster, but, Simon recalled, "Every morning he left the house with his briefcase just as all the other commuters on their way to jobs in the financial and banking districts in New York." Instead of work-related papers, "Dad had his lunch in his briefcase." Simon later recalled that he had separated emotionally from his father around age twelve, when he decided "that there was nothing there." As a child, Simon felt that his father did not enter his world "except when I did something that annoyed him." Later, as an adult, he remembered his father as a man who could not provide for his family and who tended toward loose thinking and ad hominem comments. Simon was closer to his mother. Yet he felt that she had withheld her approval of him. Looking back on his childhood, Simon remembered "little joy" and few "celebrations and happy moments." He felt closer to his two unmarried aunts, who stayed behind in Newark, and he retained a positive feeling for crowded, bustling cities through his life.[5]

Though Simon missed the cocoon of immigrant life in the city, he enjoyed some suburban pleasures, like playing hockey on the millpond and riding his bicycle. He joined the Boy Scouts and became an Eagle Scout at fourteen. Simon later recalled that he "delighted especially in the nature-study merit-badge learning. I loved building footbridges over streams using only vines and tree limbs, and I was proud of my skill at making fires." When he was twelve, Simon went away to Boy Scout camp for six weeks. He ended up in a bunk with boys from his Millburn scout council, all white Protestants who were a year older than him and camp veterans. Simon remembered them hazing him until finally he went "crazy" and dared them all to fight. The boys backed down. The experience fed Simon's sense of himself as a combative outsider. Simon saw himself as someone whose sympathies, since childhood, lay with "the struggling poor, the powerless, and those denied opportunity by circumstance." He wrote in his autobiography, "I still dislike

Julian Simon in his Boy Scout uniform, early 1940s.

Courtesy of the family of Julian Simon.

elite practices and attitudes." Though Ehrlich, growing up a few miles away under similar circumstances, hardly came from a wealthy or elite background, Simon's predilections would fuel his clashes with the environmental establishment as it gained influence and was embraced by powerful, mainstream institutions.[6]

Even as he disdained the elite, however, Julian Simon craved the external validation and opportunities for advance-

ment that elite institutions offered. In 1949, he headed off to
Harvard College on a navy ROTC scholarship. Along with Ehrlich, who enrolled at the University of Pennsylvania that same
year, Simon was part of a generation of Jewish students who
increasingly populated America's top colleges and universities
after World War II. Simon took a less direct route to academia
than Ehrlich, however, and his professional success came later.
During his college years, Simon worked variously as an encyclopedia salesman, drugstore clerk, brewery worker, grass-seed
factory employee, cab driver, and worker in a tin-can factory.
The jobs gave him a visceral understanding of hard and dirty
work. Memories of breathing in grass-seed dust on the factory
floor prompted Simon's lifelong enthusiasm for technologies
that made such tedious and unhealthy jobs obsolete. Simon
also helped support himself in college with the winnings he
took away from a regular poker game, often staying up until
dawn during his senior year. He did not suffer fools lightly, but
his close college friends thought him a terrific companion.
Simon was curious and funny. He was interested in a broad
range of people, such as his close friend Aristides Demetrios,
who became a prominent sculptor.[7]

After graduating from Harvard in 1953 with a degree in
experimental psychology, Simon spent three years as a unit officer on a naval destroyer and as an officer attached to the ma-
rines at Camp Lejeune in North Carolina. He did not see combat during the Korean War, but he did travel around the world,
usually to places not frequented by tourists, such as working
port cities. He later drew on these early impressions of developing countries to remark on how their economic conditions
had improved between the 1950s and the 1970s and 1980s.
Simon left the navy to work in advertising in New York City. In
1957, he enrolled in the business school at the University of

Julian Simon in his navy uniform with his parents, most likely at his 1953 Harvard graduation. Courtesy of the family of Julian Simon.

Chicago, where he completed an MBA and a doctorate in business economics.[8]

When Simon arrived at the University of Chicago, economists such as Milton Friedman and Friedrich A. Hayek had started to challenge the New Deal economic orthodoxy. Although Simon did his doctorate in the business school rather than the economics department, these economists—giants in their field—inspired Simon with their free-market thinking. Friedman famously argued that economic liberty yielded prosperity as well as personal freedom. In essays in the late 1940s and 1950s, Friedman had attacked government regulations—rent controls and professional licenses—and pushed for the creation of new markets in previously taboo commodities, such as pollution and elementary education. Hayek, in his World War II–era *Road to Serfdom,* warned that government control over economic decision-making led to tyranny.[9]

Although Simon never studied with them directly, he considered Friedman and Hayek kindred spirits. "You can't choose your relatives," Simon later wrote. "But one can imagine." His dream family consisted of a roster of famous theorists, some of them notable conservatives: "William James as my father, Hayek as my uncle, Milton Friedman as my older brother, Theodore Schultz as my thesis adviser, and David Hume as my idol." While Simon did not explain why he chose each member of this group, presumably he linked their ideas with his own ambitions and self-perception. His attraction to James probably lay in the philosopher's theory of pragmatism and its emphasis on "scientific loyalty to facts." Simon considered himself devoted to data and to allowing data to reveal truths that sometimes ran counter to expectations. Hume also was an empiricist who urged the rejection of every philosophical system "not founded on fact and observation." Hayek, Friedman, and

Schultz all were Nobel Prize–winning conservative economists at the University of Chicago.[10]

While at Chicago, Simon met and married a graduate student in sociology, Rita James. Rita had been a leftist activist as a teenager, part of a socialist youth group in New York City. But she had left that behind. She also now found Chicago's free-market approach liberating and exhilarating. Ayn Rand's novels emphasizing individual rights and laissez-faire capitalism inspired Rita, as did Hayek's *Road to Serfdom*. After completing their degrees, Julian and Rita Simon moved to the New York area in 1961 so that Julian could start a business and Rita could establish herself as an academic. Julian Simon created a direct-mail company that sold products like flowers and fancy coffee and tea. Simon also briefly tried selling a one-dollar booklet on how to make a will and a guide for home-brewing beer, ventures that ran afoul of federal restrictions on practicing law without a license and on marketing alcohol through the mail. These were precisely the kinds of restrictions that Milton Friedman had criticized. Looking back in his autobiography, Simon noted that the rules had changed in the intervening years: guides to writing wills and brewing alcohol were now routinely sold in bookstores. His experience in business strengthened his libertarian sentiments. He complained about the "tyranny of bureaucracy," which forced small business owners to shut down viable enterprises.

After a few years of these varied business ventures, Julian Simon grew restless and decided to write a book about how to do direct mail. He thought that an academic position would lend itself to the kind of writing and thinking that he wished to do. It was a period of tremendous expansion in higher education, and universities were rapidly adding to their faculties. Julian landed an academic job at the University of Illinois–

Urbana teaching advertising. Rita subsequently found a position in the university's sociology department after getting past nepotism rules that barred spousal hires. They bought a house on quiet Busey Avenue, within walking distance of the university. David, the first of three children, was born in Urbana in 1964, with a sister, Judith, following one year later, and a second son, Daniel, soon after. Julian and Rita kept a studious and hard-working house. Rita missed a few lectures around her children's births but quickly made it back into the classroom. The couple hired someone to take care of the kids during the day so that they both could work full-time at the university.[11]

Although family life in Urbana had its idyllic aspects, Julian Simon's move into academia coincided with a dark turn in his personal life. He slipped into a depression that plagued him for about thirteen years. He later attributed his emotional struggles to a professional incident with his mail-order business—never specified. The depression, which Simon had kept secret from all but his wife, made him miserable. "I wished for death," Simon wrote in his autobiography, "but I refrained from killing myself only because I believed that my children needed me, just as all children need their father. Endless hours every day I reviewed my faults and failures, which made me writhe in pain. I refused to let myself do the pleasurable things that my wife wisely suggested I do, because I thought that I ought to suffer." Simon's children had no idea about their father's condition. He was able to enjoy taking his young kids on bicycle rides out of town to the university's experimental farms, David riding in a basket in front, with Judith in a basket behind. When they were older, Julian liked to play basketball with David and his friends. Simon particularly found refuge in his work during these long years. Successfully completing academic writing projects provided a few bursts of pleasure outside family

Julian Simon with his son David on a family trip to California in 1972.

Courtesy of Naomi Kleitman.

life. But his commitment to work also may have prolonged his depression. Simon refused medication to treat his illness because he feared that the side effects of the drugs would impair his clarity of thought.[12]

As a scholar and author, Julian Simon liked to put a basic question to the test of data to see what might be revealed. He started with a relatively mundane subject: through economic analysis, Simon sought to determine how many library books should be available in the university library and how many could be stored in new off-site storage facilities. Simon moved on to tackle more explosive topics. In one essay, for example, he estimated how much restitution African Americans might be paid as compensation for the forced labor of their enslaved ancestors. Taking a pragmatic rather than ethically driven approach, Simon focused solely on the value of unpaid labor. He found that the amount owed to the descendants of slaves more than one hundred years after slavery depended largely on the interest rate used for the calculations: fifty-eight billion dollars if one assumed that savings grew at 3 percent per year, or five trillion dollars at the higher rate of 6 percent. As was often the case in Simon's early writing about provocative topics, his article on reparations carefully skirted political questions about whether paying reparations made sense and moral questions about whether calculating unpaid labor could fully account for the human cost. Although a Jewish economist writing about reparations would certainly have considered the parallels, he made no mention of Jewish suffering or Holocaust survivors. Simon had a not-uncommon economist's desire to solve fraught societal challenges—or, in other instances, long-standing market inefficiencies—through seemingly simple, rational application of data and analysis.[13]

Rita Simon shared her husband's interest in taking on contested topics with data and a relatively neutral tone. A few years after the couple arrived in Urbana, she revisited her radical past by inviting prominent leftist leaders to speak at the university about their experiences in the 1930s. Norman Thomas, former head of the Socialist Party, Max Shachtman, a former Trotskyite, and Earl Browder, former head of the US Communist Party, all tromped out to the university in the cornfields in Illinois to revisit the radical 1930s. Rita had moved on politically from her youth, but she knew a compelling topic. The resulting compilation, *As We Saw the Thirties,* became an important piece of intellectual history.[14]

Rita would go on over her career to write about juries, women in the professions, race relations, transracial adoption, and immigration. With each subject, Simon liked to define her questions narrowly and collect survey data about them, usually stopping short of broad conclusions about the social implications of her research. In one small but telling example, in March 1966, more than eight hundred janitors, mail messengers, and maids went on strike for three days at the University of Illinois, forming picket lines in front of university office buildings and dormitories. Rita used the strike as an opportunity for her students to conduct a telephone survey of 164 faculty members to determine their attitudes toward the dispute. She published the results, showing that most faculty members would cross the picket line, in the bulletin of the American Association of University Professors. As in many of her writings, Simon wrote in neutral tones about the conflict between workers and the university over pay, work schedules, and contract duration and avoided taking a position, either pro or con, on the strike.[15]

As a successful academic and mother of three in the 1960s, Rita Simon was at the vanguard of the women's movement. She wrote about the challenges of being a woman in academia and the problem of nepotism. In a manner common to many of her generation who fought to gain a place in higher education, she also opposed gender-based advocacy and affirmative action, embracing merit alone as her standard. Simon generally resisted getting drawn into the student-led turmoil of the 1960s. She later recalled telling protesting students that she had to cut short their meeting to get home to her children and that they would have to continue their dispute the following day. Rita Simon thrived in the sociology department at the University of Illinois, becoming its first female chair in 1968. She continued to chair the department until 1983, leaving the position only for breaks during family sabbatical trips to Israel.

Simon would serve a term as editor of the leading academic journal the *American Sociological Review* in 1978. Julian and Rita had a remarkably equitable marriage partnership, both pursuing demanding full-time academic careers. Julian had his limits as a pioneer in gender equity, however, in part due to his social awkwardness and ongoing depression; when Rita agreed to chair the sociology department, Julian insisted that Rita make clear that she didn't have a "wife" and that they would not entertain at the house.[16]

After several years in the University of Illinois advertising department, Julian Simon clashed with the department's chair, and he moved to the business school in 1966 to focus on marketing. Simon also cast about for a new area of research. He settled on the issue of population growth, starting from the assumption that population growth presented a significant social problem. In his first papers on the topic, he applied his marketing expertise to the promotion of birth control. In articles in 1968, for example, Simon recommended family planning campaigns and outlined suggestions for marketing birth control. His work found an audience among advocates. In 1969, W. Parker Mauldin of the Population Council arranged for Simon to visit India to look into possible incentive schemes to promote birth control. The Population Council also offered Simon a job to develop marketing campaigns.[17]

Simon assumed at first that rising populations posed the dire economic threat that Paul Ehrlich insisted on. Simon saw a direct tradeoff between extra income and extra children. In a 1969 article on the value of avoided births, for example, he sought to calculate the costs and benefits of family planning campaigns to justify investment in these efforts. His "point of departure," he wrote, was that "many underdeveloped coun-

tries . . . would be better off economically if their birth rates
were lower. . . . It is therefore a good thing economically to
increase the number of people who practise birth control."
Simon concluded that a country gained at least $114 per averted
birth and called contraceptive distribution programs, which
cost at most $5 per avoided birth, a "fantastic economic bar-
gain." Simon described investments in family planning and
contraception as "forty times as productive" as other development investment. With some passion, he argued that family
planning campaigns were "far too small; governments *must* do
more." Simon called on governments to pay citizens a cash
bonus to avoid having children. Although a governmental pay-
ment of up to $114 per family would be "economically rational"
in India and other underdeveloped countries, Simon thought
that much lower payments could dramatically change behavior.
In his focus on cash payments, albeit by governments, Simon
showed the influence of the market-based economics that he
learned at Chicago. Even in family planning, he thought, people make "rational decisions." They did not "reproduce like animals, without even thinking."[18]

Simon's early exhortations for action to lower birthrates,
however, soon gave way to equivocation about the dangers of a
growing population. His studies of birth control marketing and
fertility drew him to probe the dynamics of family size.
Changes in income and education influenced fertility quite differently depending on a family's previous wealth and education. Additional income encouraged more educated women to
have children. By contrast, greater income discouraged less
educated women from having more kids. Additional education
also had a differential impact. More schooling discouraged less
educated women from having children far more than women
who already were relatively highly educated. In another twist,

smaller families often used additional income to have more children, while large families usually did not. In other words, fertility did not operate according to a single formula. No one strategy, it seemed, would successfully moderate or control fertility.[19]

As Simon came to appreciate the complexity of influencing fertility rates, he grew increasingly skeptical about the comparative costs and benefits of population growth. Simon later attributed his changed view to comparative data gathered by the economists Simon Kuznets and Richard Easterlin on the relation between national economic growth and population. The historical data, Kuznets and Easterlin argued in 1967 essays, did not show that population growth had undermined economic growth. "More population means more creators and producers," Kuznets noted. Danish economist Ester Boserup's studies of agricultural growth also fed Simon's budding skepticism. Boserup found that population growth increased innovation in agriculture and spurred greater saving within an economy. Contrary to Thomas Malthus, who believed that agricultural methods set limits for population, Boserup argued the opposite: population size and density determined what kind of agriculture would be practiced and would be economically efficient.[20]

Influenced by these economists, Simon began to question prevailing ideas about the dangers of population growth. Simon's change was spiritual as well as intellectual. He later described an epiphany that he had in 1969 while attending a meeting on overseas population programs in Washington, DC. Arriving early, Simon visited the Iwo Jima memorial nearby. As he contemplated the memorial to the fallen soldiers, Simon recalled a famous eulogy given at Iwo Jima by the Jewish chaplain Roland Gittelsohn, who had bemoaned the loss of potential human talent and promise. Simon later wrote, "And then I thought, Have I gone crazy? What business do I have trying to help arrange it that fewer human beings will be born, each one of whom might be a Mozart or a Michelangelo or an Einstein— or simply a joy to his or her family and community and a person who will enjoy life?" It was soon after this change of heart that Simon found himself on the stage in Urbana, during Earth Day 1970, questioning the premise that population growth posed a scientifically proven threat to society.[21]

Simon matched his enthusiasm about adding more people to the planet with new scholarship in which he argued that population growth bolstered rather than weakened the economy. In a 1975 article about infrastructure and road construction, Simon wrote that the "ill effects" of population density are "well known" in the form of congestion and less farmland per farmer and consumer. But there were also economic returns: "higher density causes more available infrastructure per worker." More people in a given area made it easier to provide common infrastructure like roads in less developed countries, thereby providing access to markets and new supplies and technical help. Simon similarly challenged economic models that assumed that higher population growth meant that people would save less and reduce their capital investment. Following Boserup, Simon argued instead that population growth increased agricultural investment, for instance, in irrigation. In a broader assessment of the impact of population on economic growth in developing countries, Simon criticized the assumption that "population growth retards the growth of output per worker." He offered specific examples, pointing out that Europe's population grew at an unprecedented rate after 1650. While rapid population growth might not help a nation's economy, Simon's model suggested that "moderate population growth produces considerably better economic performance in the long run . . . than does a slower-growing population."[22]

Simon's scholarly research thus suggested that the effect of population growth was "complex and tenuous," and "not straightforward and clear-cut." "Any judgment," Simon wrote in 1975, "is contingent on one's values and assumptions." Values mattered because economists had to decide how to assess societal progress. Did societal welfare depend on per capita income or on the number of people that a society could support? Simon embraced a utilitarian view, arguing that "the larger the number of people who are alive, the greater the welfare." He called this a "biblical-Utilitarian welfare function" because of the Bible's injunction to be fruitful and multiply. Here Simon staked out a philosophical position directly at odds with Paul Ehrlich, who asked, "What good are six billion or more people as compared to the some three billion we have now? Shouldn't we ask what people are for?" Ehrlich dismissed the idea that "large numbers of bodies" were "more desirable than a limited number of free men." Simon rejected this on ethical as well as economic grounds. "If enabling as many people as possible to have life is taken as the purpose of the economy," Simon argued, "then what is thought by many to be our greatest contemporary problem—population growth—is seen to be a triumph rather than a disaster."[23]

Even as Julian Simon came to question the threat of population growth and to foresee a better future for a growing humanity, environmental advocates doubled down on the warnings of disaster that Paul Ehrlich had made popular with *The Population Bomb*. In 1972, an international group of industrialists, scientists, and political leaders called the Club of Rome published *The Limits to Growth: A Report for the Club of Rome's Project on the Predicament of Mankind.* The book described looming threats to humanity from population growth and excessive resource consumption. Dennis and Donella Meadows, Jørgen Randers, and William Behrens III, the book's authors, were management experts and scientists at the Massachusetts Institute of Technology. They worked there with Jay Forrester, an electrical and computer engineer who pioneered the use of models and computer simulations to analyze the dynamics of social systems. In his 1971 *World Dynamics,* Forrester had developed a model of global population, resources, and pollution that predicted an end to the economic progress of the industrial era. His young colleagues, with Dennis Meadows as the lead, developed Forrester's approach further to add layers of complexity to their simulation. Their world system's "basic behavior mode" resembled that of Ehrlich's butterflies, with "exponential growth of population and capital, followed by collapse." *The Limits to Growth* authors wrapped this doomsday perspective in a computer model that gave added weight to their predictions. The book's jacket flap warned:

> WILL THIS BE THE WORLD THAT YOUR GRANDCHILDREN WILL THANK YOU FOR? A world where industrial production has sunk to zero. Where population has suffered a catastrophic decline. Where the air, sea, and land are polluted beyond redemption. Where civilization is a distant memory.
>
> *This is the world that the computer forecasts. What is even more alarming, the collapse will not come gradually, but with awesome suddenness, with no way of stopping it.*

Although confident in their own technological prowess, the MIT modelers doubted that new technology would enable humans to circumvent natural limits. "When we introduce technological developments that successfully lift some restraint to growth or avoid some collapse," they explained, "the system simply grows to another limit, temporarily surpasses it, and falls back." The book's sponsors at the Club of Rome called for a "Copernican revolution of the mind," declaring that "only a conviction that there is no other avenue to survival can liberate the moral, intellectual and creative forces required" to achieve a state of global equilibrium.[24]

Presented to great fanfare to a crowd of more than 250 US senators, representatives, heads of agencies, business leaders, and others at the Smithsonian Institution in March 1972, *The Limits to Growth* provoked a furious public and scholarly debate. The enthusiastic response tapped the fears and anxiety made popular by Paul Ehrlich and other environmental leaders. The embrace of *The Limits to Growth* also reflected a growing doubt about economic growth as the best way to reduce poverty. The book became an international best-seller, with more than twelve million copies in print. Ehrlich himself wrote a blurb for the book, calling it "a great service"; he challenged readers unsettled by it to work "for changes in the real world." The *Christian Science Monitor* heralded *The Limits to Growth* as an "awakening trumpet" whose blast humanity must heed. "If this doesn't blow everybody's mind who can read . . . then the earth is kaput," another writer declared. *New York Times* columnist Anthony Lewis called *The Limits to Growth* "one of the most important documents of our age." Lewis embraced the book for demonstrating "the truth ecologists have been trying to teach us: that the elements of life are interconnected. They make a whole, a circle, and they have to be considered together." Limits in one area could not easily be avoided without consequences elsewhere.[25]

In a December 1972 *New York Times* magazine article inspired by *The Limits to Growth*, Rutgers University zoology professor Bertram Murray illustrated how scientists applied

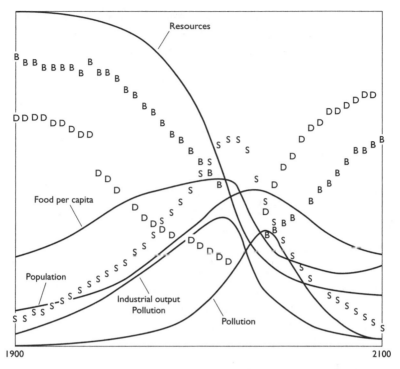

In the "World Model Standard Run," the authors of *The Limits to Growth* showed
a future with no change to existing trends. According to their model, rapidly
diminishing material resources and food supply, as well as rising pollution, would
cause a significant collapse in population by the middle of the twenty-first century.
(B stands for birth rate; D stands for death rate; and S stands for services.)
Meadows et al., *Limits to Growth*, 124.

"ecological laws" to the growth of human society and how they
believed that humans were a species like any other, subject to
the same constraints. Americans, Murray wrote, were being
told that they "must make a choice" between "a continuous
growth economic system and a no-growth economic system."
He asked rhetorically, "How are Americans to decide who is

right, the optimists or the pessimists?" Murray sided with *The Limits to Growth* and the Ehrlich-aligned pessimists. Citing "ecological principles" concerning continually growing populations, Murray argued that "there is no question that collapse is inevitable." He pointed out that ecologists and economists studied "the same phenomena." While economists eagerly sought economic growth, biologists called continuous growth "cancer" in biological tissues. Jumping from biology to political science, Murray lumped socialism and communism together with capitalism as similarly "progrowth and wasteful." A new alternative economic system, "consistent with ecological theory," was needed: "a no-growth system." This economic system would be "highly regulated" and global—"a worldwide, environmentally responsible economic system . . . managed by an international team of planners, most reasonably organized by the United Nations." Murray echoed Paul Ehrlich and John Holdren, insisting that technologists could not "get around ecological laws." Murray's article illustrated the broad public engagement by biologists in theorizing about human societies and the antagonism developing between their ecological models and prevailing economic models. His article also underscored the broad embrace by many environmentalists of national and international governance, both for environmental regulation and for economic planning. *The Limits to Growth* provided a political slogan for the early 1970s, as well as an intellectual model for thinking about how societies function.[26]

The Limits to Growth's argument that an out-of-control economic system would overshoot available resources attracted so much attention because policy-makers and commentators had already started to lament the end of an era of inexpensive energy and minerals. Ralph Lapp, a nuclear physicist, warned in

the *New York Times* in March 1972, a few weeks after the release
of *The Limits to Growth,* that America's insatiable appetite for
energy faced a natural resource supply that was "all too finite."
The United States "faces serious energy shortages in the decades ahead," Lapp wrote. "In the richest land on earth, we
have simply overwhelmed nature. . . . Our days of energy affluence are over." American oil imports had risen to record levels,
and oil companies seeking new supplies had been forced to
search for energy in remote regions like Alaska, Algeria, and
Siberia. Policy-makers at the highest level agreed. Congressional hearings in the fall of 1972 focused on the "foreign policy
implications of the energy crisis." In mid-November 1972, Secretary of Commerce Peter Peterson told thousands of oil executives gathered for the annual meeting of the American Petroleum Institute that energy would be the most important new
presidential initiative in 1973. "The era of low-cost energy is
almost dead," Peterson told the oil executives. "Popeye is running out of cheap spinach." In an influential essay in *Foreign
Affairs* the following spring, James Akins, a top State Department oil adviser and later ambassador to Saudi Arabia, chose a
different metaphor for an incipient oil crisis, declaring, "This
time the wolf is here."[27]

The wolf arrived on October 16, 1973. After the United
States sided with Israel in its October 1973 war with Egypt and
Syria, Arab oil-producing countries cut their oil production and
refused to sell petroleum to the United States. The Arab oil embargo confirmed policy-makers' worst fears about the nation's
vulnerability to the "oil weapon." Resource constraints sud-
denly threatened American prosperity. During the five-month-
long embargo, oil prices quadrupled from around three dollars
per barrel to more than twelve dollars. The sudden supply cut
and price surge inflamed fears about resource scarcity and, for

Gasoline dealers in Portland, Oregon, displayed signs explaining a flag policy
indicating who could buy gas during the fuel crisis of 1973–1974.
Courtesy of National Archives and Records Administration.

many contemporary observers, seemingly confirmed the thesis
of *The Limits to Growth*. "Running Out of Everything," declared
a *Newsweek* cover that November, depicting a fearful Uncle Sam
gazing into an empty horn of plenty. In the immediate wake of
the embargo, Congress passed laws to boost US oil production
and cut consumption. The political stalemate over the Alaskan
oil pipeline quickly ended. For years the pipeline had been held
up by lawsuits over environmental impact statements and
Alaskan native land claims. Now a pipeline bill sailed through
Congress, one that explicitly barred further judicial review over

its environmental impact. Around the same time, Nixon signed the Amtrak Improvement Act of 1973, citing the "emerging trends and international events" that had made "the energy efficiency of rail" a "national interest." In December, Congress also passed a nationwide daylight savings law and, in early January, a measure reducing the speed limit for all vehicles to fifty-five miles per hour. Nixon justified both on the basis of energy conservation. The measures marked the start of a new national campaign to reduce US dependency on foreign oil. Presidents Nixon, Ford, and Carter, as well as congressional leaders, all made energy policy a priority through the rest of the 1970s.[28]

While public anxiety about resource scarcity and environmental limits deepened in the early 1970s, many commentators and economists remained skeptical—even in the face of the oil embargo—that the world faced long-term resource scarcity and ecological limits. These critics, who included Julian Simon but went well beyond him, denounced *The Limits to Growth's* conclusions as a dangerous threat to world peace and stability. Economic policy-makers like Hollis B. Chenery, a World Bank economist, thought that curtailing economic growth would deepen poverty and inequality and lead to war and revolution. Henry Wallich, an economic expert on international finance, similarly criticized the book's antigrowth policy recommendations. In a 1972 speech to newspaper editors, Wallich warned that stopping economic growth, by limiting resource consumption or raising energy prices, would "deny the aspirations of billions of people" for a better way of life. He called such measures "suicidal" because of the social turmoil and unrest they would cause. Writing in the *Wall Street Journal,* reporter David Anderson argued for an alternative to the "absolute halt to growth" suggested by *The Limits to Growth*. Instead, Anderson said, society could achieve "selective growth" that would permit "human survival" and "continued functioning of economic life as it is now known" but lead to less waste and pollution. This alternative would involve "much higher levels of affluence, population and technology than the world knows at present, much higher than would remain if humanity somehow made some drastic attempt to halt growth now." To avoid the possible threat of scarcity, Anderson suggested, antigrowth advocates risked a greater danger, the continued impoverishment of billions of people around the world.[29]

Other critics questioned the accuracy of the *Limits to Growth* model, disputing the book's feedback loops among population growth, resources, death rates, and pollution. They particularly ridiculed the seductiveness of the new computer models. Max Lerner, writing in the *Los Angeles Times*, mocked, "You can see the curves rising, falling, galloping, dancing, converging, interlocking, but always moving toward doom." In a bit of lyrical criticism, he continued, "Ashes to ashes / And dust to dust. / If the bomb doesn't get you / The exponential curves must." More harshly, the *New York Times Book Review* called the book "little more than polemical fiction." *The Limits to Growth,* the *Times* reviewers declared, was not so much a "rediscovery of the laws of nature but . . . [a rediscovery] of the oldest maxim of computer science: Garbage In, Garbage Out." The book's computer model, they said, "takes arbitrary assumptions, shakes them up and comes out with arbitrary conclusions that have the ring of science."[30]

These critiques in the press of *The Limits to Growth* reflected skepticism especially within the community of academic economists. In the 1960s and early 1970s, as economists studied the factors leading to economic growth, they increasingly questioned the importance of natural resource inputs and instead emphasized human capital, technology, and innovation. The economists Harold Barnett and Chandler Morse, who dismissed fears of limits and resource exhaustion in their 1963 book, *Scarcity and Growth: The Economics of Natural Resource Availability*, particularly influenced Julian Simon. Barnett and Morse argued that the economics of natural resources had changed with humankind's "increased knowledge of the physical universe, changes which have built technological advance into the social processes of the modern world." "The notion of an absolute limit to natural resource availability," they wrote, "is untenable when the definition of resources changes drastically and unpredictably over time." Take the example of Vermont granite, used for generations only for construction and tombstones. With the discovery of nuclear power, Barnett and Morse argued, the uranium content of one ton of granite equaled the energy potential of 150 tons of coal. The authors slightly exaggerated this equivalence between granite and coal, since the uranium in granite is expensive and difficult to extract. But they sought to make a larger point: the scientific age "differs in kind, and not only in degree, from the preceding mechanical age." Ubiquitous materials, such as "sea water, clays, rocks, sands, and air," now became economic resources.[31]

Barnett and Morse laid out a theory about how markets interacted with technology and science to explain why discoveries such as nuclear energy were predictable consequences of human ingenuity rather than "essentially fortuitous." Scientific advances made it possible to use the abundant energy of the universe to overcome apparent constraints. "A limit may exist," Barnett and Morse conceded, "but it can be neither defined nor specified in economic terms." Modern humans had a flexible relationship to the physical world in which they lived. Scarcities and new alternatives did not necessarily imply higher costs. Particular products might increase in price, but substitutes and innovations would bring down costs or change the market. In a market society, Barnett and Morse wrote, "changes in relative costs, shifts of demand, the wish to develop broader markets— all aspects of growth—create problems which then generate solutions." In essence, they concluded, natural resource problems were "qualitative" concerns. People might choose to address population growth out of a voluntary desire to avoid a feeling of overcrowding or to protect a cherished environment, but they did not have to fear for the basic subsistence and survival of humanity.[32]

Barnett and Morse spoke for a generation of economists, including Julian Simon, who reacted skeptically to predictions of calamity. Yale economist William Nordhaus dismissed the model and calculations of Jay Forrester's *World Dynamics,* the 1971 book that fathered *The Limits to Growth,* as "measurement without data." Nordhaus said that Forrester allowed for "no technological progress, no new discovery of resources, no way of inventing substitute materials, no price system to induce the system to substitute plentiful resources for scarce resources." The model, in other words, did not match how human economies actually work. In the real economy, Nordhaus explained, abundant resources and new technologies responded to scarcity: "iron, aluminum, and communication satellites replace copper; chlorine replaces iodine; the xerography process replaces use of tin and lead in printing." This substitution would continue, Nordhaus said, unless the future turned out to be "very different from the past." Nordhaus complained that the Forrester model—and, by extension, *The Limits to Growth*— treated human society as a "population of insentient beings, unwilling and unable to check reproductive urges; unable to invent computers or birth control devices or synthetic materials; without a price system to help ration scarce goods or motivate the discovery of new ones." Human beings, Nordhaus's analysis suggested, had more options than Ehrlich's butterflies.[33]

MIT economist Robert Solow, who later won the Nobel Prize in economic sciences for his work on economic growth theory, also harshly attacked the *Limits to Growth* model and its predecessors. Solow called the studies "worthless as science and as guides to public policy." He ridiculed models that made "no room for everyday market forces." The price system allowed capitalist economies to react to relative scarcity. Markets would manage depletion by causing people to develop substitutes, use resources more efficiently, and increase production. In a prestigious 1973 Ely lecture to the American Economics Association, Solow assured his audience, somewhat glibly, that "the world has been exhausting its exhaustible resources since the first cave-man chipped a flint." Society did not need to fear crossing natural limits to growth. More important, possible future scarcity did not mean that the current generation needed to slash its consumption. "Earlier generations are entitled to draw down the pool" of natural resources, Solow insisted in an essay on intergenerational equity, "so long as they add . . . to the stock of reproducible capital." In other words, because labor and capital goods could substitute for resources, Solow wrote, natural resources should simply be consumed according to the same rules that governed other assets.[34]

Although Solow believed in market efficiency, he acknowledged that market failures also occurred in resource use, as in other areas of economic policy. Resource producers might not take all social factors into account, including pollution and waste, and they could exhaust a resource more quickly, or more slowly, than was optimal. Solow pointed to inaccurate information as a key cause of inefficient rates of resource consumption. He suggested that the government could play an important role by engaging in a "continuous program of information-gathering and dissemination" on trends in "technology, reserves, and demand." Solow suggested that taxes could be used to incorporate the social cost of extraction into private calculations.[35]

Although he acknowledged the importance of population growth, environmental degradation, and resource exhaustion, Solow emphasized that these issues were subject to standard economic analysis and did not present ominous and unmanageable threats. He argued that the *Limits to Growth* and similar models exaggerated and predetermined their grim outcomes. Solow complained that environmentalists treated their critics like decadent hedonists and thrill-seekers who did not care about the future of the planet. "One gets the notion that you favor growth if you are the sort of person whose idea of heaven is to drive at 90 miles per hour down a six-lane highway reading billboards in order to pollute the air over some crowded lake, itself polluted with the exhaust of twin 100-horsepower outboards, and whose idea of food is Cocoa Krispies." But rather than provide an important warning that might lead to sensible action, Solow concluded, the *Limits to Growth* models "divert attention from remedial public policy." He asked, "Who could pay attention to a humdrum affair like legislation to tax sulfur emissions when the date of the Apocalypse has just been announced by a computer?"[36]

Certainly not all economists were as sanguine about the future as Nordhaus and Solow, just as not all natural scientists shared Ehrlich's gloomy outlook. Robert Heilbroner, a historian of economic thought and economist at the New School for Social Research, declared the outlook for humankind "painful, difficult, perhaps desperate." World population growth portended a "grim Malthusian outcome," he wrote in his 1974 book, *An Inquiry into the Human Prospect*. Heilbroner, influenced by Ehrlich's writings on the Green Revolution, described high childhood mortality rates as "a human tragedy of immense proportions, but also a demographic safety valve of great importance." He shared Ehrlich's fear that improving childhood mortality and food production in the short term posed the "danger" of enabling "additional hundreds of millions to reach childbearing age." Heilbroner saw "only two outcomes imaginable": either the descent of the undeveloped world into social disorder or the rise of oppressive governments capable of "halting the descent into hell." Though widely respected for his accessible writings on such great political economists as Smith, Malthus, and Marx, Heilbroner, however, did not represent the mainstream of the economics profession, which was increasingly technical and focused on quantitative modeling. Other economists toward the edge of the field, such as Ehrlich's intellectual ally Herman Daly, sought to create a new environmental or ecological economics that could challenge mainstream assumptions about the importance of economic growth. Daly and his colleagues rejected the central premise of Solow, Nordhaus, and other mainline economists: that economic models, and the economic markets, could adequately account for the costs of pollution and economic growth, and that labor and capital could largely substitute for nature.[37]

During the early 1970s, Julian Simon wrote largely about technical aspects of fertility and population, as well as other economic issues. He published in academic journals of economics, demography, and development. His work did not drive the public debate over *The Limits to Growth* and the energy crisis. Simon's general perspective on population and resource scarcity instead reflected increasingly common ideas among economists such as Solow and Nordhaus about the adaptability of market economies. These economists thought about scarcity much differently than ecologists did. Where ecologists viewed scarce resources as a fundamental constraint that either forced dramatic change or provoked crisis, economists saw scarcity and abundance as constantly shifting, dynamic variables. In an ecosystem, scientists argued, resource abundance prompted excessive growth, followed by collapse when scarcity returned. In a marketplace, economists said, abundance also yielded growth; scarcity, however, did not lead to crisis. Rather, scarcity created challenges that spurred the economy in new directions. Even as ecologists moved toward a greater awareness of interdependence in ecosystems and the unique roles that individual species played, many economists grew increasingly detached from biological systems and argued against natural constraints.[38]

Julian Simon's strident attacks on prevailing ideas about population growth, starting with his remarks around the first Earth Day in 1970, went further rhetorically than many other economists cared to follow. Simon shocked many by declaring that population growth should "thrill rather than frighten us." He stood apart from other economists in his willingness to repudiate completely the orthodoxy on population growth. Simon's faith in the infallibility of markets also exceeded that of more mainstream economists like Nordhaus and Solow. Liberals and environmentalists, such as the zoologist Paul Silverman, Simon's University of Illinois antagonist, viewed Simon as a simple-minded ideologue with a Panglossian attitude toward grave environmental problems. The tension between these two perspectives, the optimists and the pessimists, deepened during the 1970s. Environmentalists gained influence over national policy-making, particularly under President Jimmy Carter. Meanwhile Julian Simon and other critics sharpened their attack, developing a full-throated critique of population control and environmental regulation.[39]

CHAPTER THREE **Listening to Cassandra**

I n June 1974, Paul Ehrlich and his close friend John Holdren flew from California to Washington, DC, to testify before Congress. They had been invited to share their thoughts on a proposal to require government economists to gather and share data on key commodities and materials. The Senate hearing revealed how much *The Limits to Growth* and fears of scarcity gripped the imagination of Congress in the wake of the 1973 oil embargo, which quadrupled oil prices and created the sense of a national energy crisis. Washington senator Warren Magnuson, who chaired the hearing, declared that the "choking off" of critical raw materials and energy resources was "at the heart of our economic sickness." Senator Charles Percy, a moderate Republican from Illinois, similarly described the oil embargo as an "agonizing and humiliating experience" that the nation now risked with other raw materials. Economists who disagreed, Percy said, were just "whistling in the dark." The senators believed the nation was at the end of an era of natural abundance. Walter Huddleston, Democrat from Kentucky, described "a collision course between nature's supply and man's demand. . . . Neither this planet, this particular continent, nor this country can any longer be viewed as an eternal fountain from which we can forever drink."[1]

"Resource experts throughout history have become a chorus of Cassandras," complained Senator Gaylord Nelson of Wisconsin, a leading environmentalist in Congress who had proposed the first Earth Day celebrations and teach-ins. Nelson referred to the Greek prophetess whose warnings about future catastrophes, though accurate, were ignored. Like Cassandra, scientists like Ehrlich and Holdren possessed the ability to predict "worldwide catastrophe" from population growth and resource scarcity but faced the curse of not being believed. The entire world now risked the "dire consequences of Cassandra's predicament," Nelson warned.[2]

The enthusiastic embrace of Ehrlich and Holdren by leading politicians indicated how environmentalists were outgrowing the Cassandra role—their message, in fact, was widely heard and increasingly influential. Ehrlich and Holdren shared prophecies of scarcity and devastation at the Senate hearing. "A new era in the world" was coming, Ehrlich said, taking people from an "age of abundance to an age of scarcity." Within a decade, food and water would become difficult to obtain and expensive, resulting in a billion or more people starving to death. Supplies of key minerals would near depletion, depriving industries of critical materials. The "growthmanic economy" would plunge into depression, and Americans would sink into a no-growth or negative-growth economy.[3]

Speaking after Ehrlich, Holdren defended *The Limits to Growth, The Population Bomb,* and similar books. Critics of *The Limits to Growth* argued that market prices, technological innovation, and inexpensive energy could "somehow bail us out." Holdren disagreed. He thought that economists like Robert

Solow and William Nordhaus misunderstood the role of their own discipline. Economics, Holdren acknowledged, helped allocate scarce resources efficiently through prices. But prices and markets could not make "scarce resources less scarce." Holdren thus returned again to the idea of fundamental ecological limits that economics could not overcome. Economists understood resources, labor, and capital as interchangeable factors in economic production, but Holdren insisted that ecosystems could not be maintained by artificial means. No "technological equivalent" existed for natural ecosystems. Humans could not run Earth like an "Apollo capsule," he said, referring to the space program. He saw "unwarranted technological optimism" as the "most dangerous tendency" facing society.[4]

Holdren declared himself "firmly in the neo-Malthusian camp," convinced that population growth would outstrip available resources. Technological optimists faced the burden of proof, he said. "Are we worse off if we believe the pessimists and they are wrong, or are we worse off if we believe the optimists and they are wrong? I think the conclusion is clear." He echoed the conclusion of *The Population Bomb,* in which Ehrlich, claiming similarly that risk management favored environmental caution, had written, "If I'm wrong, people will still be better fed, better housed, and happier." Holdren's and Ehrlich's confidence in the virtue of their proposals reflected the way many environmentalists dismissed the counter-call for economic growth. Ehrlich and Holdren spoke freely and easily of a "no-growth" society and "de-development." They did not concede that halting growth might block people from getting out of poverty or have significant consequences of any kind.[5]

The testimony in Washington reflected Paul Ehrlich's broader message during the 1970s about the end of growth economics and the coming period of austerity. In books and speeches, Ehrlich depicted the United States as overdeveloped, consuming the planet's resources at dangerous and unsustainable levels. "What are the prospects for the future?" Paul and Anne Ehrlich wrote in their 1974 book, *The End of Affluence*. Their book criticized America's mass-consumption society along the same lines as economist John Kenneth Galbraith's popular 1958 book, *The Affluent Society*. But where Galbraith called for public investment strategies that could strengthen the economy and address social inequalities, the Ehrlichs considered the American model simply doomed. "We are facing, within the next three decades, the disintegration of an unstable world of nation-states infected with growthmania," the Ehrlichs wrote. "The game of unlimited growth is ending, like it or not. We are approaching the limits." The Ehrlichs argued, in their typical rhetorical style, that society should make voluntary changes today to avoid having future changes imposed by catastrophe. They dismissed the short-term costs of their proposals, including unemployment resulting from a halt to economic growth, as greatly preferable to the economic disaster waiting around the corner.[6]

The Ehrlichs' message went well beyond population numbers to far-reaching social transformation. Los Angeles, with its snarled traffic and urban sprawl, represented all that the Ehrlichs opposed, the "perfect model of the kind of place not to live in during an age of scarcity." City centers should be rebuilt and suburbanization halted. The Ehrlichs proposed changing the "entire pattern of transportation in the United States," abandoning federal highway construction for railroad development. Buildings and appliances should have efficiency standards, wasteful lighting and heating technologies should be banned or limited, and recycling should be expanded. Some of these ideas became central tenets of the environmental movement, for instance the idea of fostering development in city centers near transportation hubs or setting federal efficiency standards for appliances. Others gained no traction. The Ehrlichs proposed eliminating Monday holidays to encourage more local recreation and reduce driving on long weekends. They knew this was unlikely to happen, however, because the highway lobby, labor unions, and other business interests stood in the way.[7]

Given the likely failure of governments and businesses to respond to the environmental crisis—the United States would surely "blow it on the energy front"—the Ehrlichs urged individual action to reduce energy use and prepare for the "end of affluence." Some of their suggestions reflected the culture of self-denial and frugality that coexisted uneasily with the "me" decade of narcissism and budding consumption. Turn off the lights, lower the thermostat in the winter, and raise it in the summer. Air dry your clothes when possible, wear a sweater. Other suggestions reflected the growing interest in dropping out of society and achieving self-sufficiency. The Ehrlichs' suggestion that readers reduce their dependence on society verged on survivalism. Learn basic subsistence skills, such as how to find water and food. "Talents useful when stranded far from civilization might prove equally useful if society breaks down." Start to produce your own food with a garden. Stock up on food and water to be able to survive for several weeks, even a year. "If you have even a small yard, there is no reason not to keep a couple of water-filled trashcans available at all times." Buy long-lasting clothing—not just "fad fashion." Add locks to your doors and consider getting a dog for protection, although that would add another mouth to feed. Gardening and survival preparation were not activities that Anne and Paul spent much time on in their own frenetic rush of research and travel. But the Ehrlichs encouraged young people thinking of dropping out of society to get a head start on living off the grid. "Change your lifestyle now and avoid the rush later," they wrote. Every individual with a farm or with survival supplies already in hand would be one fewer person lining up for them when the crisis hit. It was time to "loosen the tightest bonds of interdependence," they concluded.[8]

The Ehrlichs urged a "relaxed lifestyle, good friends, and a happy sex life" over fame and profit. Their sentiments meshed with other critiques of American society during the 1970s that rejected materialism, global markets, and advanced technologies. The Kentucky poet Wendell Berry, in a widely circulated essay called "Think Little," argued that war, racial oppression, and pollution all were interrelated—caused by the "mentality of greed and exploitation." Berry said that all Americans bore responsibility for the environmental crisis and needed to act in both public and private ways to address it. "Nearly every one of us, nearly every day of his life, is contributing directly to the ruin of this planet." Rather than "think big," Berry called on Americans to "think little," and to drive less, consume less, and waste less. The remedies for an American character weakened by overconsumption, militarism, and waste, Berry argued, "require a new kind of life—harder, more laborious, poorer in luxuries and gadgets, but also, I am certain, richer in meaning and more abundant in real pleasure." In his 1973 *Small Is Beautiful: Economics as If People Mattered*, the British author and economist E. F. Schumacher argued similarly that societies should emphasize local, self-reliant economies developed at an appropriate scale using ecologically and socially appropriate technologies. Schumacher's book circulated widely in the United States, and "small is beautiful" became a common phrase. While some Americans simply cut back their use of technology like air conditioners, a smaller fringe of the countercultural movement actually went "back to the land." Hundreds of followers of the spiritual leader Stephen Gaskin, for instance, joined a bus caravan out of San Francisco in 1971 to settle on a commune in rural Tennessee. On "The Farm," Gaskin's followers, like many other commune dwellers, rejected materialism and individualism to embark on a new way of living more appropriate to an age of limits.[9]

Meanwhile, in 1970s Washington, lawyers and policymakers did not uproot and remake society to the extent called for by visionaries like Ehrlich, Berry, and Gaskin. But lawmakers, spurred on by environmental thinkers and advocates, did create a new legal framework for economic activity in the United States. Congress passed major environmental legislation addressing a raft of environmental issues—pesticides, ocean dumping, water pollution, drinking water, endangered species, solid waste, and toxic substances. The Environmental Protection Agency banned the pesticide DDT, whose impact Ehrlich had studied many years before as a graduate student. Unleaded gasoline, starting in 1974, and catalytic converters, beginning in 1975, sharply reduced automotive air pollution. Newly created environmental law organizations, including the Natural Resources Defense Council, Center for Law in the Public Interest, and Sierra Club Legal Defense Fund, aggressively pushed the courts to force implementation of the new federal laws. Landmark Supreme Court cases involving the trans-Alaska pipeline, nuclear power plants, and a proposed ski resort near Sequoia National Park in California successfully blocked or slowed major development projects. A radical pro-environment sentiment even penetrated the Supreme Court, where Justice William O. Douglas argued passionately in the ski resort case that inanimate objects should have a kind of legal standing: "The river as plaintiff speaks for the ecological

unit of life that is part of it. Those people who have a meaningful relation to that body of water—whether it be a fisherman, a
canoeist, a zoologist, or a logger—must be able to speak for the
values which the river represents and which are threatened
with destruction." By the mid-1970s, with remarkable speed,
many of Paul and Anne Ehrlich's environmental ideas had become mainstream, represented at the highest levels of American power, including the White House.[10]

On Thursday, December 12, 1974, Jimmy Carter, the little-
known governor of Georgia, went to the National Press Club in
Washington to announce his candidacy for president. Carter
focused much of his speech on post-Watergate political reforms, including measures to promote transparency in government and reduce corruption and the influence of special interests. But as if taking a cue from Ehrlich, Carter also warned of
the complex threats posed by "increases in world population,
food shortages, environmental deterioration, [and] depletion
of irreplaceable commodities." Americans needed to change
their wasteful ways and plan the rational management of energy and natural resources. "We are grossly wasting our energy
resources and other precious raw materials as though their
supply was infinite." Carter complained that the United States
"now has no understandable national purpose, no clearly defined goals." He called the nation to decisive action and moral
resolve. Quoting the Bible, Carter asked, "If the trumpet give
an uncertain sound, who shall prepare himself to the battle?"
Carter sounded a lot like Paul Ehrlich and the authors of
The Limits to Growth—a true believer in impending ecological
doom. Grave dangers lay ahead if the United States did not
curb its wasteful consumption and prepare for a new era of
limits. "We must even face the prospect of changing our basic

President Jimmy Carter pulling on his wading boots, August 1978, Grand Tetons, Wyoming. Courtesy Jimmy Carter Library.

ways of living," Carter said. "This change will either be made on our own initiative in a planned and rational way, or forced on us with chaos and suffering by the inexorable laws of nature." His comments echoed the arguments of ecologists that the "laws of nature" would come down harshly on humanity, not unlike the wrath of God.[11]

Carter's love of nature and his awareness of its constraints and limits drew from his earliest years growing up on a farm in a small town in central Georgia. Carter worked hard on the farm, handling farm animals, collecting eggs, feeding chickens, and helping with plowing and planting. He and his friends also played in the woods and swamps, hunting squirrels, rabbits, and other small game and fishing for eels and catfish. Carter continued this close engagement with the natural world as an adult. As governor of Georgia in the early 1970s, Jimmy and his wife, Rosalynn, took weekend trips around the state, where they "rode the wild rivers in rafts, canoes, and kayaks." Carter particularly loved Cumberland Island in southeast Georgia, where sea turtles laid their eggs in early summer. "We would watch the sun rise over the Atlantic," Carter later recalled, "and drive down twenty miles of the broad white beach without seeing another living soul." Carter's embrace of state conservation policy drew on his fear that rapid land development was laying waste to his beloved Georgia landscape and that historic landmarks were being "destroyed by bulldozers."[12]

Carter's passion for nature provided one outlet for the moral fervor that infused his life and helped shape the political biography that he shared with the nation as he ran for president. Carter was a born-again Christian. He did not smoke, and he drank sparingly. He attended church regularly as a deacon and occasionally preached and did traveling missionary work before being elected governor. "I don't know how to compromise on any principle I believe is right," he wrote in his 1975 autobiography. His rejection of compromise was certainly an exaggeration for this successful politician, but it also reflected an important truth. Carter's firm sense of principles infused his political language and decision-making with moral righteousness. He spoke in judgment-laden language that criticized the irresponsible despoiling of the environment. "Avarice, selfishness, procrastination and neglect" threatened Georgia's "great natural beauty and promise," he said in a speech as governor. Carter continued to hammer home this theme after he became president, calling the United States the "most wasteful nation on Earth."[13]

In addition to his love for the outdoors and his moralizing philosophy, Carter's faith in regional planning drew him to natural resource policy. A nuclear engineer by training and the son of an avid agricultural improver, Carter believed in the power of science and expertise to address social problems. In the years before his election as Georgia's governor, Carter built his statewide reputation through government planning. After his defeat in the 1966 Georgia gubernatorial primary, Carter spent the next four years laying the groundwork to run again. He helped to organize an eight-county planning and development commission and served as its chairman for several years. He also helped start a new statewide planning society to help support the regional planning commissions. Carter's interest in planning carried over into his term as governor, starting in 1971. He held more than fifty public meetings around the state to engage thousands of Georgians in discussions about "long-range goals in public life." Carter thought that government should plan for contingencies twenty years out and address the "interrelationships among societal factors." The planning initiative underlay Carter's efforts to rationalize state government, which, he boasted, led to the abolishment and consolidation of 278 of 300 state agencies.[14]

Carter also believed in family planning, which was partly stimulated by his mother's experiences in India in the late 1960s. His mother, Lillian Carter, was a registered nurse who signed up for the Peace Corps in 1966 to work on family planning issues. As Carter described it in his autobiography *Living Faith*, Lillian Carter wrote on her application that she wished to be assigned to "a place where people are dark-skinned and desperately in need." She was sent to a small village near Bombay (now Mumbai) to work on a national birth control and family planning initiative. She worked with a program that involved persuading men to get vasectomies through incentives and threatened loss of social services. One afternoon she helped with thirty-three vasectomies. Jimmy Carter rejected these coercive approaches to limiting family size or curbing population growth and instead favored voluntary planning efforts. As an evangelical Baptist, Carter strongly opposed abortion, seeing it as the taking of a human life. Georgia passed a strict anti-abortion law while he was governor, and Carter opposed federal funding for abortion. He generally supported the *Roe v. Wade* decision establishing a constitutional right to abortion, however, and backed family planning efforts more broadly.[15]

Carter's concern for nature thus fused with his moral fervor about limits and sacrifice and his faith in planning to make him a political embodiment of many aspects of the 1970s liberal sensibility. Carter used his environmental interests to build his national political profile, for example by joining a United Nations committee on the environment. Carter attended the June 1972 UN Conference on the Human Environment in Stockholm, where, in a side auditorium, Paul Ehrlich held his rhetorical duel with Barry Commoner's allies over the significance of population growth. With the exception of his efforts to reorganize Georgia's state administration, Carter said that he "spent more time preserving our natural resources" as governor than on any other issue. As governor, Carter called for energy efficiency, recycling, energy planning and research, and raising the thermostats to reduce air conditioning. As early as 1973, he warned that the United States faced an "energy crisis" and he called for a "Manhattan Project" on energy to develop a broad array of fuels. Carter's grand vision for government-led industrial development revealed a split within the environmental movement over whether technology offered solutions as well as threats. Carter envisioned large-scale technical solutions to ecological problems, not just the small-scale alternatives promoted by people like E. F. Schumacher. As Carter prepared in 1974 to launch a presidential run, he called for a coherent national vision and direction. He believed that the people and their leaders could harness their resources to "devise effective, understandable and practical goals and policies in every realm of public life." Addressing the problems of resource scarcity and population growth figured prominently on the list of Carter's global concerns.[16]

Newly elected president Jimmy Carter took the podium on 20 January 1977 to deliver an inaugural speech to a nation disillusioned by Watergate and the Vietnam War. The day was cold and sunny, with a wind-chill temperature in the teens. Many Americans were ready for a president who could provide moral leadership to the United States and help the nation plan for a coming age of scarcity. "'More' is not necessarily 'better,'" Carter told the country. "Even our great Nation has its recognized limits." Carter gave notice that the *Limits to Growth* mentality now had a champion in the White House. Solar panels powered the reviewing stand for the inauguration. To show their humility and their commitment to energy efficiency and simplicity, the new president and his wife, Rosalyn, broke with tradition by walking the mile and a half down Pennsylvania Avenue from the Capitol to the White House. Carter eliminated chauffeur service for his White House aides to save money. In his first fireside chat, the president wore a warm sweater in keeping with his theme of energy conservation and frugality.[17]

Within days of Carter's presidential inauguration, a natural gas shortage thrust energy policy to front-page news, creating both a crisis and an opportunity for the new president. "Blizzard tightens energy crisis," blared the front-page headlines of the *New York Times*. "This year, cry of gas shortage 'wolf' is real," declared the *Baltimore Sun*. An unusually cold winter in the eastern United States had boosted demand for natural gas. At

the same time, government controls governing the price of natural gas in interstate pipelines meant that producers refused to
ship more of their gas supply east at the lower regulated prices.
They could get far higher prices keeping their gas out of the
interstate trade. As January 1977 ended, some nine thousand
factories and business shut down temporarily, and, according
to the White House, half a million workers had lost their jobs
because of gas shortages. Another 1.6 million jobs were said to
be at risk. School districts kept hundreds of thousands of
schoolchildren home from school because of a lack of natural
gas for school buildings. Carter declared Pennsylvania and
New York federal disaster areas because of the combination of
cold weather and shortages of heating fuel. The president was
photographed with his hands outstretched as he slipped on the
ice—it was an apt metaphor for the way that nature had in-
truded to shake up the first days of Carter's presidency.[18]

Although the natural gas crisis resulted largely from a dysfunctional, regulated market, Carter thought that the gas shortages represented the future of energy scarcity that he had
talked about during his campaign. In his first presidential trip,
Carter flew by helicopter to Pittsburgh, where he visited a par-
tially shut-down Westinghouse factory. Wearing what he described as "heavy long underwear" and a warm sweater, Carter
warned that the nation faced a "permanent, very serious energy
shortage" that would require a "comprehensive national energy
policy." "The crisis might be over in a few days or a couple of
weeks, but the energy shortage is going to be with us, is going
to get worse instead of better." The president promised to deliver a comprehensive energy bill by 20 April, ninety days after
his inauguration. In the meantime, Carter asked Congress for
emergency legislation to allow the federal government to divert

natural gas supplies to the northeast and to temporarily lift some price controls to encourage more sales of gas through the interstate system.[19]

The natural gas crisis in his first weeks in office brought Carter and the energy issue together in a way that would define his presidency. Carter had already asked James Schlesinger, a longtime Washington hand with a reputation for shaking up bureaucracy, to formulate a comprehensive national energy policy. Schlesinger was an economist and national security expert who had been secretary of defense under Presidents Nixon and Ford, director of the Central Intelligence Agency, and chairman of the US Atomic Energy Commission. Brought in over the objections of environmentalists who disliked his past support for nuclear power, Schlesinger was part of Carter's effort to create a bipartisan administration. Schlesinger later recalled that the natural gas crisis "pulled energy problems to center stage. . . . This was an area in which [Carter] was getting some things done and it appealed to his moral sense, it appealed to his sense of history. It appealed to his sense as an engineer." Energy policy thus united the different aspects of Carter's political identity—his moral purpose, his faith in planning, and his belief in the inexorable constraints of nature.[20]

In his response to the natural gas shortage, Carter differed from his predecessors in his passion for energy conservation. Conservation meant more to the president than simply a strategy for addressing resource supply. Carter was "a moralizer in regard to energy," said Schlesinger. "He just thought that we were too damn wasteful." Carter made conservation a cornerstone of his energy program, Schlesinger said, out of a "moral conviction that we should be provident in our use of the resources that have been placed by the Almighty on this earth." In the middle of the natural gas crisis, the president ordered federal buildings to lower their thermostats to 65 degrees during the day and 55 degrees at night. The drafty and cold White House would serve as a national example for homeowners and building managers. As White House custodians set about turning down the thermostats, Carter reportedly told the first meeting of his National Security Council, "This is the last warm meeting we'll have."[21]

Not everyone in the White House agreed willingly to this personal sacrifice. Rosalynn Carter remembered her shock, having recently arrived from Georgia: "I couldn't believe it: I had been freezing ever since we moved in. My offices were so cold I couldn't concentrate, and my staff was typing with gloves on. . . . I pleaded with Jimmy to set the thermostats at 68 degrees, but it didn't do any good." Rosalynn had to resort to long underwear and slacks and get used to the cold winter. Vice President Walter Mondale later recalled that Carter had the air conditioners turned off in the summer as well. "I remember the hot Washington summers of 1977 and 1978—it would be ninety-five degrees outside, and Carter would have the White House air-conditioning turned off. We would sit through cabinet meetings and it would be hotter than a bug and muggier than hell—all of us wiping our brows. He would sit right through it because he wanted to provide an example for the country." White House maintenance staff—nicknamed the "thermostat police"—would periodically visit offices to make sure that staffers hadn't reset the thermostats. National Security Advisor Zbigniew Brzezinski reportedly moved a lamp near the thermostat in his office so that the heat would cause his air conditioning to turn on.[22]

In April 1977, Carter said it was time for an "unpleasant talk" with the American people about energy. He called the "energy crisis" an unprecedented problem and, other than war, "the greatest challenge that our country will face during our lifetime." He embraced the limits-to-growth attitude toward natural resources. The world was "simply running out" of oil and natural gas, and people could use up "all the proven reserves of oil in the entire world by the end of the next decade." Carter warned that the "energy crisis" would worsen for the remainder of the century and threatened to overwhelm the nation. Americans should not to be "selfish or timid." The nation simply had to balance its energy needs with its shrinking resources. To address the problem, he called for "unpopular" proposals that would demand "sacrifices" and cause "inconveniences."[23]

Carter acknowledged that the immediate urgency of the 1973 oil crisis had disappeared but insisted that the nation's energy problem had actually worsened. Domestic oil production continued to drop, and more time had passed without a plan for the future. Carter predicted that by the early 1980s, the world would demand more oil than it could produce. He bluntly stated that the United States was running out of oil and must embrace strict conservation measures and increased use of coal and other energy sources. He painted a picture of a United States humbled and weakened by its own inaction. By failing to plan for the nation's energy future, Carter said, the United States "will feel mounting pressure to plunder the environment. We will have to have a crash program to build more nuclear plants, strip mine and burn more coal, and drill more offshore wells. . . . Inflation will soar; production will go down; people will lose their jobs. . . . If we fail to act soon, we will face an economic, social, and political crisis that will threaten our free institutions." Facing "national catastrophe," this test of the American character would be the "moral equivalent of war."[24]

Carter's energy speech presented an apocalyptic vision equal to Ehrlich's predictions of mass devastation from famine and *The Limits to Growth* calculations of overshoot. Yet he differed from Ehrlich in having the opportunity, and the burden, of translating his vision into policy. The task would consume the rest of his presidency. In his April speech, Carter laid out the principles for a national energy plan, one that would offer something to almost every domestic source of energy production or efficiency. Carter promised to cut back oil use by expanding consumption of coal and unconventional energy sources like oil shale and solar power while also reducing demand through energy conservation. To encourage domestic exploration and development, Carter called for removing some oil price controls that Nixon had imposed in 1971 to fight inflation. A new Department of Energy would provide cabinet-level urgency to federal energy policy. In a November 1977 speech to the nation, Carter reiterated that the situation only worsened "with every passing month." He warned that overdependence on foreign oil risked unemployment, trade imbalances, and national security. Carter pointed to an "unpleasant fact" about energy prices: "They are going up, whether we pass an energy program or not, as fuel becomes more scarce and more expensive to produce." Carter called his energy plan a "good insurance policy for the future."[25]

House Speaker Tip O'Neill pushed Carter's package through intact in five months. But the plan fell apart in the Senate. Proposals to move away from federal regulation of the energy markets—representing a profound break with decades-old New Deal–era economic thinking—met fierce resistance. Many liberal Democrats in Congress opposed decontrol of oil and natural gas prices. They feared that price increases would boost industry profits without benefiting consumers. Oil state Democrats, such as Louisiana senator Russell Long, chairman of the Senate Finance Committee, in turn blocked new taxes

President Jimmy Carter speaking at the dedication of the White House solar panels in 1979. © Harvey Georges/AP/Corbis.

proposed for oil that might offset "windfall" gains that might result from price decontrol. Natural gas producers complained that price controls unfairly cut into their profits and stalled development.[26]

Carter had modest success in formulating a new energy

policy, but the lengthy struggle over his top domestic priority undermined his political strength and his broader policy agenda. Carter successfully established a new Department of Energy to coordinate federal energy administration, and he expanded funding for nonpetroleum fuels. Carter also pushed through mandatory efficiency improvements. Yet he failed to get congressional support for tax policies that might have raised the cost of energy. In the end, the struggle over oil and natural gas pricing, government controls, and taxation tied up too much of Carter's political capital. Carter and liberal Democrats in Congress held onto the idea that oil and gas were critical strategic resources whose prices and production levels had to be managed through government controls. Stuart Eizenstat, Carter's domestic policy adviser, later described the bitter fight over price deregulation, particularly for natural gas, as a "tragic error." Carter "never recovered" from the natural gas fight "in terms of public perception of him as a President." In his memoirs, Carter compared working on energy policy to "chewing on a rock that lasted the whole four years."[27]

As Jimmy Carter struggled to translate concerns about scarcity and limits into a politically viable energy policy, Paul Ehrlich and Julian Simon continued to develop their starkly different analyses of population growth and natural resources. In 1977, Paul and Anne Ehrlich and John Holdren jointly published a thousand-page college textbook entitled *Ecoscience: Population, Resources, and Environment*, which dramatically expanded and revised their earlier publications on the "predicament of humanity." Noting energy price shocks, famine in Africa and India, and the spread of nuclear weapons, the authors said that these problems were "interlocked" with economic distress and rapid population growth. Humanity faced a "grave crisis." On a positive note, environmental awareness had spread widely. "Everyone who is alertable is alerted." Now effective action depended on understanding complex relationships between human and natural systems.[28]

Ecoscience, with its sophisticated and integrated approach to human societies, ecosystems, and natural resources, demonstrated how much scientific understanding had increased in the post–World War II years. After sketching the fundamental principles of geology and ecology, the authors analyzed the demands that growing human populations placed on food production, forests, water, and energy. They showed how humanity depended on, and dangerously threatened, ecological systems vital to society. The authors also documented "direct assaults" on human well-being from air and water pollution and dangerous chemicals.[29]

The Ehrlichs and Holdren concluded that humanity's current path was simply "impractical *physically*." Calculating absolute limits—the millions of metric tons of fish that could possibly be extracted from the ocean, for example—they rejected any possibility that innovation and markets could continue to expand those production frontiers. The authors quoted Fairfield Osborn's 1948 book *Our Plundered Planet:* "The tide of the earth's population is rising, the reservoir of the earth's living resources is falling." Food production could not keep up with population growth. Agricultural development programs would only buy "badly needed time" necessary to control the population explosion. If *Ecoscience*'s arguments had a distinctive and characteristic weakness, it lay in the authors' supreme confidence that they could calculate the absolute limits of human and environmental potential. One British ecologist remarked in a generally positive review that *Ecoscience* expressed the "ecological evangelism" of the previous decade while largely ignoring the contributions that economics might make. Although marketed as a textbook, *Ecoscience* called for action to halt population growth, including financial incentives for smaller families, birth control, and possible "compulsory control of family size."[30]

As Ehrlich and his coauthors considered—without endorsing—inflammatory ideas such as marketable licenses for children and mass sterilization campaigns, Ehrlich found himself forced to devote more and more time to addressing the backlash against his positions. Critics continued to denounce population control, and Ehrlich personally, as racist and anti-poor for focusing on controlling births and immigration. One of Ehrlich's scientific colleagues at Stanford, the Nobel Prize–winning physicist William Shockley, spoke disparagingly of the intelligence of African Americans and called for eugenic policies to strengthen the genetic evolution of the human species. Ehrlich despised Shockley and his views on race, but he also saw that critics often linked his own campaign against population growth with the eugenic ideas of people like Shockley.[31]

Ehrlich sought to demonstrate that population and immigration control efforts rested on a scientific foundation that could be kept separate and untarnished by racism and anti-immigrant sentiments. During the mid-1970s, Ehrlich partly turned his attention from biological research to socially fraught issues of race and immigration. In 1977, the same year he published *Ecoscience*, Ehrlich also coauthored *The Race Bomb: Skin Color, Prejudice, and Intelligence*, with Stanford psychologist Shirley Feldman. Ehrlich and Feldman's book attacked Shockley's ideas about "race," saying that the "race-IQ debate" was a "scientifically useless discussion." "Since there are no biological races to begin with," Ehrlich and Feldman wrote, "the question of the inferiority or superiority of a race is meaningless."

Yet they recognized the powerful impact of racial ideas, which were helping to "push civilization toward the brink of catastrophe." "Few things," the authors wrote, "stand more squarely in the path of extracting mankind from its present predicament than the complex of problems commonly associated with skin color and lumped under the heading of 'race.'" "The problems of race relations were difficult enough in an era when it seemed that the pie was ever expanding and that sooner or later there would be abundance for everyone. Now that this vision of abundance is fading fast, the probabilities of racial conflict may well be increasing." Ehrlich and Feldman gestured at riots and racial conflicts in Watts in 1965, Boston in 1974, and Soweto in 1976, calling them possible "harbingers of much worse to come." They argued, contrary to most historical interpretations, that these social conflicts reflected the stresses of overpopulation and conflict over limited resources.[32]

Ehrlich and Feldman sought to save population and family planning programs from the taint of racism. They acknowledged that some people "would like to see programs of population control initiated to reduce the numbers of 'inferior' people." But they insisted that the US government provided birth control information and services "to benefit the poor," not to carry out "genocide" against African Americans. Ehrlich and Feldman quoted W. E. B. DuBois endorsing birth control and Martin Luther King Jr. on the importance of family planning in reducing unwanted children. They argued that preventing the provision of birth control and family planning services was in fact "acting genocidally" because the "underprivileged of any nation will suffer first and most from their own increased numbers."[33]

From race, Paul Ehrlich turned to immigration. He now roamed far afield from biology and butterflies. Just as in *Race*

Bomb, where Ehrlich sought to dispose of the "nonsense" surrounding race and population control, now he aimed to reshape the immigration debate to make it politically safe to call for strict limits on immigration. Ehrlich's new focus on immigration reflected changing population dynamics in the United States. Domestic fertility rates—the average number of children for women of reproductive age in a given year—had fallen dramatically since the mid-1950s. From a baby-boom peak in total fertility of 3.78 in 1957, the rate fell to 2.46 in 1969, just after Ehrlich published *The Population Bomb*. The total fertility rate then dropped further in 1975 to 1.74, below what would be thought of as replacement level. Ehrlich's advocacy for smaller families, as well as expanded access to birth control and legalized abortion, may have played a small part bringing down the fertility rate after 1968, but the rapid movement of women into the workforce and changing family economics were more significant. At the same time, the US immigrant population continued to grow and to change in its composition. In 1965, Lyndon Johnson had signed the Immigration and Nationality Act, which abolished the restrictive system that had excluded Asian and African immigrants, as well as southern and eastern European ones. Illegal immigration from Mexico and elsewhere in Latin America also was increasing. Between 1970 and 1980, the foreign-born population of the United States increased by 46 percent, from 9.6 million to more than 14 million. Ninety-six percent of that increase came from Asia or Latin America.[34]

Ehrlich and other population control advocates shifted their attention from American family size to immigration as a key factor in US population growth. Many mixed concern about population growth with fears that immigrants took work from native-born Americans. As one Zero Population Growth spokesperson wrote in the *Washington Post* in 1974, "At a time when we are having great difficulty providing jobs for our own native-born citizens, employment competition by immigrants is an increasingly serious problem." In 1977, Zero Population Growth launched a national campaign to press for curbs on legal and illegal immigration. A fundraising appeal signed by Paul Ehrlich described illegal immigration as a "human tidal wave" that was "depressing our economy and costing American taxpayers an estimated $10 billion to $13 billion a year in lost earnings and taxes, in welfare benefits and public services." In June 1977, Zero Population Growth's board proposed cutting annual legal immigration from 400,000 to 150,000 people per year and stopping illegal immigration. From an alliance with liberals over women's rights and abortion, population control advocates switched to common cause with the anti-immigrant political right.[35]

When Ehrlich came under attack for being anti-immigrant, he struggled to defend himself by drawing nuanced distinctions between his positions and those of "bigoted" Americans. In 1979, Paul and Anne Ehrlich, together with a historian of migration named Loy Bilderback, published *The Golden Door: International Migration, Mexico, and the United States*. In the book, Paul Ehrlich distinguished himself forcefully from those who thought that illegal immigrants were "shiftless and lazy." He now rejected the criticisms of immigrants that he and organizations like Zero Population Growth had advanced earlier in the 1970s. "All responsible studies indicate that the illegals work diligently for modest wages. By and large, they do work that no one else is willing to do. Contrary to the folklore, as far as we can tell, they pay their taxes and seldom apply for or receive public assistance." Ehrlich similarly dismissed the "simplistic view that each illegal displaces an American," calling it "utter nonsense." He lamented the "failure of Americans to understand and appreciate Mexican culture and history" and argued that the flow of immigrants historically had been "governed very largely by American self-interest, racial prejudice, and political expediency." Most of the public concern over a Mexican immigration "crisis," Ehrlich contended, could be "traced directly to scare tactics of bigots and bureaucrats."[36]

But Ehrlich was not really a liberal on immigration. He wanted not more open borders but rather to establish a non-racist, scientifically justified rationale for setting strict limits. As he had the urban riots of the 1960s and 1970s, Ehrlich saw immigration largely through the lens of population growth and overpopulation. He called migrants coming to the United States for economic opportunity "demographic refugees" forced out by overpopulation. Even if Americans bore responsibility for their historically complex relationship with Mexico, Ehrlich wrote, the United States could not serve as a "safety valve" for Mexico's population growth. Achieving a balanced relationship with nature meant not just conservation and modifications to the American way of life but also fewer Americans. "The kind of life that most Americans expect puts practical limits on many things, including population size. And however desirable each individual immigrant may be, he or she increases the number of Americans." Ehrlich called for an "*explicit* population policy" that would further lower American birthrates and allow for "rational planning." Ehrlich insisted that racism and xenophobia could be set aside. Immigration policy, he thought, should be solely a question of desired population level. Ehrlich concluded *The Golden Door* with a pitch and contact information for two of his favorite organizations, Zero Population Growth and the Federation for American Immigration Reform. Ehrlich called the immigration group "based in the population-control movement" and said that the organization was "dedicated to developing restrictionist policies that are humane and consistent with modern democratic values."[37]

While Ehrlich in the late 1970s pressed for population and immigration controls consistent with his social views, Julian Simon became more strident and public after years of quiet scholarly work at the University of Illinois. In his 1977 book *Economics of Population Growth,* Simon attacked the rationale for government population controls and denounced Paul Ehrlich specifically. Did population growth really threaten either the environment or the economy? Simon described the recent surge in population as a key "sign of success of human civilization." "The fact that our economic system can support an ever-increasing number of people seems wonderful to me," he wrote. Simon called Ehrlich "morally abhorrent" for his stated opposition to medical efforts to reduce mortality, what Ehrlich called mockingly "short-sighted programs of death control." Simon lamented that distinguished scientists and professors such as Ehrlich inflamed popular fears with loaded terms like "population explosion," "people pollution," "population bomb," and "the population plague."[38]

Simon's skepticism about physical limits to population or economic growth sharpened to an attack on "popular Malthusian belief." Simon complained privately to natural resource economist Harold Barnett in 1977, "Don't you get terribly frustrated seeing the kinds of arguments that are brought to bear in national discussions of natural resources and energy?" Pointing to the immense power of the Sun, Simon asked Barnett, "Is there any reason to think that energy as we define it and understand it is limited?" For all practical purposes, Simon wrote, "there are no ultimate limits." Simon's belief in limitless resources seemed "to argue against all logic," he acknowledged in the *Economics of Population Growth*, but Simon insisted that "commonsense is just plain wrong." "Commonsense notices our use of resources but fails to see that our needs lead to our creation of resources—planting of forests, exploration of new oil fields and invention of ways to obtain oil from rocks, discovery of substitute sources of energy and nutrients, invention of new tools of all kinds. Clearly we now have available to us vastly more resources of almost every kind than did people in any preceding age." Resources, Simon thought, were becoming more abundant, not scarcer.[39]

The sense of thrill and exhilaration that Julian Simon started to feel about population growth carried over to his personal life. He finally broke the grip of the debilitating depression that had darkened his life for over a decade. Through a form of cognitive therapy in which he attacked his own tendency to make negative self-comparisons, Simon claimed to have overcome his depression in just a few weeks. Starting in April 1975, he recalled, "I have almost always been glad to be alive, and I have taken pleasure in my days. I have occasionally even been ecstatic, skipping and jumping from joy." Simon later would write a guide to cognitive therapy to try to share his success with others. Cognitive therapy was another way that human innovation could solve a seemingly natural problem, in this case one of mental well-being.[40]

In addition to cognitive therapy, Simon also identified his cure for his depression with his new practice of observing the Jewish Sabbath every week. "The Jewish Sabbath is the center of our family's week and we do whatever we can to make it serene and joyful," Simon wrote. From Friday night through Saturday evening, Simon would set aside all work and try to put aside negative thoughts and speech. Instead, he would spend the day relaxing, reading history and other nonfiction books, talking on the telephone, and visiting with friends and family. Simon was not a deeply religious person. In 1975, for instance, he took his kids to see basketball player Kareem Abdul-Jabbar play in an exhibition match on Kol Nidre, the night of Yom Kippur and perhaps the holiest day of the Jewish year. At family Friday night dinners, he enjoyed talking about the insights of Buddhism. In a personal note for his 1978 Harvard reunion, Simon wrote, "Because the Buddha was the purest atheist of all, I'm a Buddhist by theology." At the same time, Julian and Rita cherished the weekly break that observing the Sabbath gave them. Rita and Julian also enjoyed a deep commitment to Israel and the "glory of Jerusalem." They spent several sabbatical years living there with their children. During one sabbatical, Julian Simon regularly joined the security patrols in their Jerusalem neighborhood. Rita Simon grew so close to their Moroccan housekeeper in Jerusalem that she eventually set up a small foundation in the housekeeper's name to provide academic scholarships to North African Jews in Israel.[41]

Back in the United States, Julian Simon's professional life improved, and he gained new recognition and success. Simon had a significant personal triumph when his quirky proposal for an airline-bumping auction was adopted in 1978 by the federal agency that oversaw air travel. In 1968, responding to the problem of passengers showing up for flights only to find that their seat had been given away, Simon had advocated that airlines pay willing passengers to give up their oversold seats voluntarily. He pressed intermittently for his idea against resistant airlines and skeptical economists. Simon's strategy finally found a receptive ear at the Civil Aeronautics Board, the federal agency responsible in the 1970s for regulating competition in passenger air travel. Jimmy Carter had started to deregulate some longtime heavily regulated industries, such as airlines.

Carter ordered agencies to assess the comparative costs and benefits of new initiatives and created a regulatory review council for new regulations. Cornell economist Alfred Kahn, Carter's appointee to chair the Civil Aeronautics Board in 1977, led the board to adopt Julian Simon's recommendation on airline overbooking, saying that "its appeal to an economist is obvious." Airlines would now offer payments to volunteers who would give up their seats when a flight had been oversold. The new system allowed airlines to sell more tickets for the same number of seats, thereby raising profits, while also reducing the number of people who were bumped involuntarily. Simon considered it a personal triumph, exclaiming to a colleague, "It is rare that an academic has an opportunity to influence the government process, with an idea, even a little bit." Soon after the plan's adoption, Simon wrote to free-market economist Milton Friedman, whom Simon had cultivated as a correspondent and mentor. Friedman had doubted the plan's feasibility. Simon boasted, "I'm not given to I-told-you-so's, but . . . I told you so." Simon called his victory a clear example of how "markets can improve life for all concerned parties." By 1991, six hundred thousand people per year voluntarily accepted compensation in exchange for switching to a later flight. It was one of Simon's more practical accomplishments and illustrated for him the power of markets to improve social welfare.[42]

Simon also made his first forays into immigration policy, contending that immigrants should be viewed as assets rather than a liability. In 1979 testimony before the Select Commission on Immigration and Refugee Policy, Simon argued that immigrants adjust quickly to life in the United States. Rather than being a financial burden on society, they actually contributed relatively more than native-born residents. Because immigrants generally moved to the United States when young, they

paid taxes for many years before drawing significantly on society's social services. Simon said that immigrant workers
brought a 40 percent return on society's investment in them.
Ehrlich and Simon agreed that immigrants themselves were
not a liability to society. They disagreed, however, on whether
the United States should accept more people into the country.
Ehrlich urged cuts to immigration into the United States to
reduce the nation's population and its environmental impact.
Simon thought immigration should be encouraged as a spur
to economic growth.[43]

As the 1970s came to an end, Ehrlich and Simon drew
closer to each other's orbit. Simon fixed his sights directly on
Ehrlich as a chief antagonist. Ehrlich's expansive and sharp
rhetoric made him a natural foil for Simon, who also was prone
to bluster himself. In 1980, Simon would launch a blistering
attack on Ehrlich in both academic and popular forums. Ehrlich, however, had not yet noticed Simon as a particular threat;
indeed, he never would acknowledge the unconventional economist as a worthy opponent. Throughout his career, Ehrlich
dismissed anti-environmental arguments as easily refuted by
science. He suggested that his critics were hucksters selling
defective goods.[44]

Yet by the late 1970s, Ehrlich had recognized that much as
he liked to treat people like Simon as fools best ignored, they
threatened his message. The final summary chapter of *Ecoscience* contrasted two views of the future: "cornucopian" and
"neo-Malthusian." Ehrlich and his coauthors criticized "cornucopian" thinkers for presuming that cheap energy and technological innovation could generate widespread abundance and
for underestimating how extensively proposed technologies
would degrade the natural environment. Ehrlich argued that

the loss of what he called "ecosystem services," such as soil fertility or climatic balance, posed the "gravest threat to human well-being."[45]

Although once a marginal view within American society, Ehrlich's frankly "pessimistic view of the human predicament" had become mainstream at the close of the 1970s, with powerful adherents at the highest branches of the US government. Shortly after taking office, the Carter administration embraced the *Limits to Growth* computer modeling approach by trying to create an official government version. Carter's Council on Environmental Quality hired Gerald Barney to develop a national model to project trends in energy, population, and natural resources through the year 2000. Barney had spent a postdoctoral year at MIT in 1970, working closely with the systems dynamics group as it prepared the *Limits* project. Barney then worked on projects related to the *Limits* study while at the Nixon administration's Council on Environmental Quality. He then went to the Rockefeller Brothers Fund, where he made grants to *Limits* coauthor Donella Meadows and to MIT for continued systems work. He also collaborated with Donella Meadows on the 1977 edited volume *The Unfinished Agenda: The Citizen's Policy Guide to Environmental Issues*, which warned of growing scarcity of resources and recommended that all forms of food and agricultural assistance "be linked to bringing birth rates into line with death rates." In his concluding chapter, Barney emphasized the challenges of the "world transition from abundance to scarcity." This transition involved more than just "physical limits to growth"; it also encompassed a "profound change in values" that would shift the emphasis from "personal interests and individualism" to the "interests of the whole society." Reflecting a typical New Age spirituality, Barney quoted Robert Pirsig's *Zen and the Art of Motorcycle*

Maintenance on the importance of aesthetic quality and Aldo Leopold on the Land Ethic.[46]

Barney sought to use his report to President Carter as a means to institutionalize systems modeling and forecasting in the federal government. Barney's team therefore based all of its predictions on internally generated government statistics. Before Barney's effort, the federal government lacked a global model for resources, population, and the economy. Instead, government agencies used isolated sector-specific models that did not interact with each other or work in a dynamic fashion. Along the lines of *The Limits to Growth,* Barney and his colleagues argued that cumulative impacts and feedback loops would determine future carrying capacity and ecological stability. "Without appropriate feedbacks," Barney later argued, previous models were "biased toward unjustified optimism." Feedback loops, of course, could work in either direction, favoring optimism or pessimism, depending on the specifications of the model.[47]

Barney's team—which included Anne Ehrlich as a consultant—produced a report in six months that criticized the federal government's models and predicted a dim future. The report echoed *The Population Bomb* and *The Limits to Growth,* warning that "if present trends continue, the world in 2000 will be more crowded, more polluted, less stable ecologically, and more vulnerable to disruption than the world we live in now." Some critics within the White House, particularly Carter's domestic policy advisers, feared the political implications of the report. They questioned whether the government should embrace Cassandra-like warnings so fully. After two years of delay, the White House finally released *The Global 2000 Report to the President* in July 1980. *The Global 2000 Report* ended up becoming a global phenomenon, selling 1.5 million copies in nine languages. Food and energy prices would more than double by 2000, the study predicted. Poverty and hunger would "haunt the globe," according to news coverage of the report.[48]

As Carter's first term neared its close, the pessimism of *The Global 2000 Report* matched the mood of many Americans. The national economy remained stalled. Inflation ran between 11 percent in 1979 and 18 percent in 1980, and unemployment rates were over 7 percent. In November 1979, Iranian radicals seized control of the United States embassy in Tehran, taking dozens of Americans hostage. Carter's inability to resolve the Iranian hostage crisis during his presidency underscored a growing sense of powerlessness and paralysis in the nation. A terrifying accident at the Three Mile Island nuclear power plant in 1979 raised new questions about the safety of atomic power and the fallibility of technology and human expertise. Scientists also warned about the possibility of nuclear winter in the event of a nuclear war with the Soviet Union. Many Americans felt that events were spinning out of control. They feared for the future but disagreed sharply on which path forward promised to return national prosperity.

Jimmy Carter called on Americans to "conserve energy" and "eliminate waste" and warned of "material limits." Carter had listened to Cassandra. He devoted his precious political capital to trying to pass energy legislation premised in part on the idea that the United States was quickly "running out of petroleum." Carter's strategies didn't simply involve conservation. He funded energy efficiency and solar energy, but he also called for the expanded use of coal and he bolstered the US military commitment in the Persian Gulf. Underlying it all lay Carter's belief that demand for oil would outstrip production in the early 1980s. If action wasn't taken, he said in one speech, the world's entire proven reserves of oil could be gone by 1990.

Viewed through a later lens, Carter's fears about looming short-ages appear exaggerated. A glut of oil drove world oil prices down by around 80 percent (in constant dollars) from their height in 1980 to a subsequent low in 1998. But listening to Cassandra also meant that Carter brought a new level of federal support for energy efficiency and alternative fuels, which have become increasingly vital elements of our energy system and share few of the environmental and social risks associated with petroleum. Yet Carter's mixture of moralism, emphasis on limits, and fear of scarcity did not appeal to all Americans. The question for the country at the end of his presidency was this: Had Carter outlined the principled and necessary path toward environmental progress? Or was he holding the United States back and unnecessarily constraining its future?[49]

CHAPTER FOUR   The Triumph of Optimism

T he path to Julian Simon and Paul Ehrlich's bet led through the intellectual jousting of scholarly journals and newspaper op-ed pages. Simon and Ehrlich clashed in print directly for the first time in the summer of 1980. In the June issue of *Science*, Simon launched a blistering attack on environmental doomsayers. He opened the article by debunking a *Newsweek* and United Nations story that more than a hundred thousand West Africans had died of hunger caused by drought between 1968 and 1973. In fact, only a small fraction of that number had died as a result of the drought. Exaggerated statistics were an all-too-common tool of manipulation, Simon argued: bad news about population growth, resources, and the environment "published widely in the face of contradictory evidence." Simon similarly questioned estimates that arable land was disappearing. Rather than more farmers working smaller plots of land to eke out subsistence, Simon said, fewer farmers produced more food and fed more people than ever before, particularly in higher-income, industrialized countries.[1]

At the center of his attack, Simon put Ehrlich's *Population*

Bomb and other warnings of population-driven famine. Ehrlich had suggested in his book that limited food supplies might necessitate compulsory population control and a form of national triage that would cut desperate countries off from food aid. Yet, Simon pointed out, food supply had increased 25 percent during the previous quarter century. Farmers in the United States in 1980 worried about "disaster from too much food." Deaths from famine globally had decreased since World War II.

Simon also slammed the idea—popularized by *The Limits to Growth*—that natural resources are finite and humanity approached ecological limits. This "apparently self-evident proposition," Simon wrote, was actually "downright misleading." Energy was "getting more plentiful," not scarcer. Using the example of copper, Simon made his own extreme claims about mineral resource abundance. He rejected the idea that copper supplies would ever run out. More copper could be made from other metals, he said. "Even the total weight of the earth is not a theoretical limit to the amount of copper that might be available to earthlings in the future. Only the total weight of the universe . . . would be such a theoretical limit." With these claims about the infinitude of available copper verging on alchemy, Simon pushed his ideology to its limits. But his essential argument reflected basic economic thinking about the advance of technology and substitution of different materials. "Because we find new lodes, invent better production methods, and discover new substitutes," he wrote, only the limits of human knowledge constrained "our capacity to enjoy unlimited raw materials at acceptable prices."[2]

In conclusion, Simon asked why "false statements of bad news" dominated public discussion. He blamed financial incentives for researchers who sought grant funding and the fact that "bad news sells books, newspapers, and magazines." Simon also suggested a psychological explanation, arguing that people tended to compare the present and future with an "ideal state of affairs" rather than with the past. More careful comparisons with past trends would reveal steady improvement in human welfare, Simon thought. New exponential growth models, such as those used in *The Limits to Growth*, tended to "seduce and bewitch" their users. And all of this had the proverbial effect of the boy who cried wolf. Rather than "harmless exaggeration," apocalyptic predictions by Ehrlich and other environmentalists, Simon thought, resulted in a "lack of credibility for real threats" and a "loss of public trust."[3]

Simon's *Science* article, which he also excerpted for the *Washington Post* opinion page, infuriated the Ehrlich camp. *Science* published a flurry of bitter letters from them in December 1980. Paul and Anne Ehrlich, along with their close colleagues John Holdren and John Harte, jointly denounced Simon's article as full of "striking misconceptions." Ehrlich and his colleagues insisted that energy and mineral scarcity was a real and present threat. They called Simon's idea that copper could be made from other metals "preposterous." They attacked Simon for suggesting that it was proper to "appropriate all the earth's resources" to support human beings. They argued that technology could not replace services provided by ecosystems to regulate climate, water cycles, solar radiation, and other essential processes. The scientists derided Simon's "tired old argument" as typical of economists who "know nothing about geology." Wayne Davis, an expert on bat migration and biology at the University of Kentucky—no geologist himself—insisted that minerals and fossil fuels were scarce. Simon's prediction that oil prices would continue to fall "defies logic," Davis scoffed.[4]

In response to the letters, Simon dismissed the idea that a new era of resource scarcity had begun, one that marked a "discontinuity" with long-term resource trends. He believed the economic forces that had yielded progress in the past would continue to spur innovation and market solutions in the future. Simon also insisted that he had not said that "all is well everywhere." The future was not simply "rosy." "Children are hungry and sick; people live out lives of physical or intellectual poverty, and lack of opportunity; war or some new pollution may finish us." Yet it did not help the world's poor to insist that things were getting worse instead of recognizing the improvement in aggregate economic trends. Simon also acknowledged that he simply differed from the scientists on the basic question of the "rights of nonhuman species to exist." "In tradeoffs between human beings and the rest of nature," Simon wrote, "my sympathies usually lie with people."[5]

Ehrlich and Simon's rhetorical battle continued into the spring of 1981, spilling into the pages of the *Social Science Quarterly*. "How often does a prophet have to be wrong before we no longer believe that he or she is a true prophet?" Simon goaded. He argued that Ehrlich had been wrong about the "demographic facts of the 1970s," whereas Simon's own predictions had been right. Ehrlich had said in 1969, for instance, "If I were a gambler, I would take even money that England will not exist in the year 2000." Ehrlich had been expressing his view that, without worldwide population control, overpopulation would cause nuclear war, plague, ecological catastrophe, or disastrous resource scarcities. Complaining that Ehrlich made wild statements without ever facing the "consequences of being wrong," Simon said, "I'll put my money where my mouth is" and asked Ehrlich to do the same. Rather than betting on the future existence of England, Simon challenged Ehrlich to bet on raw material prices and test their theories about future abundance. Ehrlich's warnings about limits to economic growth, famines, and declining food harvests suggested rising prices that reflected growing scarcity due to population growth. But Simon argued that prices generally were falling for natural resources because they were becoming less scarce due to increasing productivity and human ingenuity.[6]

Ehrlich took the bait, accepting Simon's "astonishing offer before other greedy people jump in." Ehrlich consulted with his friends John Holdren and John Harte to choose the raw materials whose supply they thought would come under the greatest pressure. They chose five key metals. Each played a critical role in the modern economy. Chromium was a crucial element in stainless steel and valued as a corrosion-resistant coating. Copper had been used for thousands of years for its malleability and then later for its ability to conduct heat and electricity. Nickel helped make stainless steel and batteries and magnets. Tin yielded corrosion-resistant alloys. Tungsten's heat-resistant characteristics found uses in lightbulbs, cathode-ray tubes, heating elements, and alloys. Each metal had seen dramatic production increases during the twentieth century. More than 95 percent of the copper ever mined in the history of the world, for example, was produced during the twentieth century.[7]

The market price for every one of these five metals had risen by at least 59 percent (copper) and as much as 357 percent (chromium) during the 1970s, giving Ehrlich plenty of reason to believe in their upward trajectory. But because inflation ran so high during the decade—averaging more than 7 percent—the general impression of rapidly rising prices was also misleading. Adjusted for inflation, copper prices actually fell by 15 percent in real terms from 1970 to 1979. Chromium prices had still more than doubled in real terms, rising by 143 percent, but the increase was still much less than it seemed. Adjusted for inflation, nickel, tungsten, and tin rose 11 percent, 76 percent, and 126 percent, respectively. At the same time, the value of the dollar also declined during the 1970s, raising prices for commodities traded on international markets.[8]

Ehrlich, Holdren, and Harte knew about inflation and exchange rates, but soaring nominal prices could not help but encourage their belief that resources were rapidly getting scarcer. Many shared their conviction. The story of Bunker and Herbert Hunt, scions of a leading Texas oil family, might have provided a cautionary tale for the scientists. The Hunts gambled billions of dollars on the rising price of silver. When prices did not increase sufficiently, the Hunt brothers tried to corner the silver market; at one point, they and their partners controlled 77 percent of the silver in private hands. Their effort failed spectacularly in March 1980, however, when government regulators tightened credit and restricted silver purchases. As silver prices collapsed, the Hunt brothers in desperation were forced to borrow more than a billion dollars to extricate themselves from their silver play. Despite such stories from the business pages, Ehrlich and his colleagues believed that the price trends all were in their favor. They felt confident that they would prevail in the bet.[9]

Ehrlich and Simon's bet, which Holdren and Harte joined, would run for ten years, through 1990, covering a thousand dollars' worth of the five minerals (a two-hundred-dollar contract for each mineral). If the mineral prices went up, adjusting for inflation, Simon would pay the difference; if the prices went down, Ehrlich and his colleagues would pay the difference to Simon. Ehrlich, Harte, and Holdren particularly liked the structure of the bet, since the value of the thousand-dollar bundle of minerals could increase without limit yet the scientists could lose no more than their thousand dollars. It seemed a small price to pay to silence Julian Simon for ten years, they thought.

For both sides, the real winnings would be bragging rights and the chance to prove that they were right about the future course of history. It was, as the *Chronicle of Higher Education* reported, "the scholarly wager of the decade."[10]

As Ehrlich and Simon worked themselves into this bet about resource prices and the consequences of population growth, Americans faced their own gamble about the future in 1980. A great deal more was directly at stake: Jimmy Carter or Ronald Reagan. Government planner versus free marketeer. Pessimist versus optimist. Cold houses and sweaters versus warm homes fueled by new nuclear power plants and oil wells. Of course, as complex political figures, neither Carter nor Reagan conformed precisely to these neat boxes. Carter, for example, had helped initiate the loosening of federal regulation of sectors of the economy such as air travel and energy. But Ehrlich and Simon's bet over mineral prices captured in miniature the clash between the two ways of thinking that seemed to frame the Carter-Reagan contest. Ehrlich and other environmental leaders helped build a powerful movement in the 1970s. But they also fueled a backlash against liberals and environmentalists that former California governor Ronald Reagan exploited in his campaign for the White House. In retrospect, Reagan and the Republican Party's extreme rhetorical turn against environmentalism in the early 1980s can be seen in part as a response to the equally extreme warnings about imminent doom emanating from Carter and environmentalists like Ehrlich.

Announcing his run for president in New York City in 1979, Reagan offered a vision of hope and limitless American growth. An American, Reagan declared, was a person who "lives in anticipation of the future because he knows it will be a great place." Reagan displayed a faith in human ingenuity that matched Julian Simon's. "Nothing is impossible," Reagan said. "Man is capable of improving his circumstances beyond what we are told is fact." Reagan derided Carter's apparent pessimism about the American future. "They tell us we must learn to live with less, and teach our children that their lives will be less full and prosperous than ours have been. . . . I don't believe that. And, I don't believe you do either. That is why I am seeking the presidency. I cannot and will not stand by and see this great country destroy itself." Reagan mocked Carter's approach to natural resource management as an "utter fiasco." The federal government had "overspent, overestimated, and over-regulated" across the board, Reagan said, and that included Carter's energy policies. "It is no program simply to say 'use less energy.'" "At best," Reagan declared, energy conservation "means we will run out of energy a little more slowly." To meet the nation's energy needs, Reagan insisted, America simply needed "more domestic production of oil and gas."[11]

Reagan's optimism about American abundance and prosperity reflected his mixed environmental record as governor of California from 1967 to 1975. Reagan, much like President Nixon at the national level, had responded to growing calls for environmental protection. Reagan called pollution a "national disgrace" that threatened the "delicate balance of ecology." He supported the creation of a state department of environmental protection and he signed forceful legislation to combat air and water pollution. Reagan also backed the creation of Redwood National Park in northern California and blocked new dams proposed for the Feather and Eel Rivers. He created an interstate regional planning authority to manage development around Lake Tahoe. He led a highly publicized pack trip into the Minarets wilderness area of the eastern Sierras, where he declared his

Ronald Reagan on horseback at his California ranch, April 1986.

Courtesy Ronald Reagan Library

opposition to a proposed trans-Sierra highway that would have broken up continuous swathes of protected land. The land preservation efforts fit well into Reagan's celebratory appreciation of the western landscape. He felt a strong affinity for the idea of the American West and for rugged outdoorsmen, and he adopted the role of the western cowboy in his political persona. Reagan loved to ride horses and owned a series of ranches, including a 688-acre property in the Santa Ynez Mountains northwest of Santa Barbara, California, which became his western presidential retreat. He enjoyed spending a "pleasant evening" with a "stack of horse and western magazines."[12]

As he signed off on some environmental measures, however, Governor Reagan viewed other proposals as unwarranted expansions of state regulatory power that threatened business development and local governance. Environmental protection, Reagan warned, should not bring "economic development to a sudden and catastrophic halt." He opposed legislation to create state coastal and energy commissions, and he vetoed funds for coastal conservation. He criticized "panic about overbuilding" in California and castigated environmentalists for "doomsday" predictions. Reagan particularly dismissed new warnings about crises, overpopulation, and starvation as "simplistic overstatement." "We used to have problems," Reagan said dismissively in a 1971 speech to the American Petroleum Institute. "Today we have crises." People like Paul Ehrlich were simply "anti-technology" and "anti-industry." "The doomsday crowd," Reagan said, "always seem to ignore the very real progress we have made." During the 1973–1974 oil embargo, Reagan said that America's abundant resources and technological prowess could make the nation energy self-sufficient. Markets and technology would resolve population problems, Reagan thought.[13]

Reagan similarly rejected the models and expert pronouncements that lay at the heart of *The Limits to Growth* and *The Global 2000 Report*. Reagan accused Carter of siding with "elitist" social planners willing to accept slower economic growth. "The limits-to-growth people who are so influential in the Carter administration are telling us, in effect, that the American economic pie is shrinking, that we all have to settle for a smaller slice." Reagan called instead for "government to get out of the way while the rest of us make a bigger pie so that everybody can have a bigger slice." In announcing his candidacy in November 1979, Reagan attacked "unknown, unidentifiable experts," such as the authors of *The Limits to Growth*, who used computer models to concoct warnings of forced scarcity. When the Carter White House released *The Global 2000 Report* at the height of the 1980 campaign, Reagan rejected its warnings about overpopulation and resource scarcity. "Well, you know there was a fella named Malthus who thought we were going to run out of food," Reagan declared in September 1980. "But Malthus didn't know about fertilizers and pesticides." While Carter's colleagues considered *The Global 2000 Report* as "significant as the Declaration of Independence," Reagan dismissed it as unfounded pessimism and flawed reasoning. Reagan insisted that resource limits were not real and should not constrain America's future.[14]

Reagan cast the 1980 election as a choice based on Americans' assessment of the state of the nation and the causes of economic stagnation. When voters went into the polling booths, Reagan said in his one debate with Jimmy Carter, they should ask themselves, "Are you better off than you were four years ago?" Reagan argued that federal regulation and planning had made Americans worse off. "This country doesn't have to be in the shape that it is in. We do not have to go on sharing in scarcity. . . . All of this can be cured and all of it can be solved." Reagan called for getting the federal government "off the people's backs" so that it would stop telling "us how to run our lives." Reagan promised to reduce the federal role and leave other responsibilities to the states.[15]

Environmental leaders, meanwhile, at first did not fully support Carter. They were disappointed in his presidency. Despite Carter's strong environmental bona fides, he had failed environmentalists on several fronts. Carter had allowed completion of the controversial Tellico Dam in Tennessee, and had loosened some clean air and water regulations. He also had supported synthetic fuels, which environmentalists viewed skeptically. These actions led some environmental activists initially to support Senator Edward Kennedy of Massachusetts in the Democratic primary. Others, like John Harte, flirted with independent candidate John Anderson. Carter's advisers tried to patch things up with environmentalists leading into the election. On the recommendation of the heads of the environmental organizations, Carter appointed James Gustave Speth, a cofounder of the Natural Resources Defense Council, to chair the Council on Environmental Quality. "Environmentalists are important to me," Carter scrawled on an August 1979 request to his chief of staff, Hamilton Jordan, to increase the profile and White House role of the Council on Environmental Quality. "We need to try to keep our environmental constituency," Stuart Eizenstat, Carter's domestic policy adviser, told a colleague.[16]

In September 1980, as the election approached, twenty-two national environmental leaders went to the White House to announce their formal endorsement of Carter. They characterized the 1980 election as a "basic choice" between Reagan and Carter. Reagan's "ignorance of environmental issues" was "bad news, really bad news," according to Marion Edey, executive director of the League of Conservation Voters. The president of the National Audubon Society, Russell Peterson, declared that the "choice is very clear to those who care about their children and grandchildren." Peterson's endorsement was especially notable because he was a former Republican governor of Delaware and had served as chairman of the Council on Environmental Quality under Presidents Nixon and Ford. Peterson thought Carter was "facing up" to the long-term problems facing the planet, while Reagan displayed a "basic misunderstanding" and determination simply to "free up industry so that it can make a bundle today."[17]

Reagan's proposed expansion of energy production revealed the great difference between the two candidates. Carter had made energy efficiency and renewable energy central to his plan to reduce oil imports. He distrusted the international oil companies and supported additional taxes that would prevent the companies from capturing "windfall profits." Reagan, by contrast, thought that the United States could solve its energy problems if "the government would get out of the way and let the oil companies explore and drill and produce the oil we have." Reagan mocked Carter: "They say, 'Turn down the thermostats, drive less, or don't drive at all.'" America just needed to "set the oil industry loose." Large quantities of oil and gas lay beneath the land and offshore, but Carter discouraged their development. Coal and nuclear power also had the potential to supply energy to millions of homes but were thwarted by "obstructionist campaigns." "I am an environmentalist," Reagan said during the campaign, but he thought that the Environmental Protection Agency tended to "insist on unreasonable and many times untried standards." Reagan linked environmental concerns to national economic growth, declaring that the "economic prosperity of our people is a fundamental part of our environment." If the "no growth" officials at EPA had their way, Reagan complained, "you and I would have to live in rabbit holes or birds' nests." He horrified environmentalists and editorial page writers by claiming, incorrectly, that the May 1980 volcanic eruption at Mount St. Helens and the decomposition of plants and trees released more pollutants than automobiles and power plants. Reagan thought that the nation's air pollution had largely been cleared up since 1970 and that regulatory standards "helped force factories to shut down and cost workers their jobs." Reagan embraced the Nevada-based Sagebrush Rebellion, which attacked federal land management and sought to shift control of federal lands to the states and private owners to encourage more rapid development. Environmental leaders came to view Reagan's defeat as an urgent priority.[18]

Yet in November 1980, Reagan won the presidency with more than 50 percent of the popular vote and more than 90 percent of the Electoral College. Carter claimed a handful of states and 41 percent of the popular vote, with Independent candidate John Anderson garnering 6.6 percent. Many factors led to Carter's rout. The economy remained mired in a toxic mix of high inflation, high interest rates, and unemployment, making it difficult for any incumbent president to be reelected. The Iranian hostage crisis and the Soviet Union's invasion of Afghanistan elevated international disputes and raised questions about the forcefulness and effectiveness of Carter's foreign policy. In addition to these factors, however, Reagan's victory also meant defeat for Carter's vocal embrace of limits and his warnings about the future. Environmental advocates in the 1970s, and Carter himself, sought broad public support for constraints on growth and reduced consumption. But many Americans resisted calls to change their behavior. The majority instead voted for Reagan and his faith in an abundant future and his skeptical view of government.

After his defeat, Carter continued to press for action on environmental and population issues. In his farewell address in January 1981, Carter emphasized the themes of *The Global 2000 Report* as key elements of his legacy. Citing "real and growing dangers" to the air, water, and land, Carter warned of the "rapid depletion of irreplaceable minerals, the erosion of topsoil, the destruction of beauty, the blight of pollution, the demands of increasing billions of people . . . problems which are easy to observe and predict but difficult to resolve." According to one of his speechwriters, Carter spent more time shaping and rewriting the farewell address than any other speech of his presidency. Carter hoped to sustain momentum around *The Global 2000 Report* and to pressure the Reagan administration to address the problems it identified. "If we do not act," Carter said, "the world of the year 2000 will be much less able to sustain life than it is now." In subsequent comments, Carter insisted that *The Global 2000 Report* was "not a prophecy of doom; it was an expression of confidence and hope—provided warnings were heeded and appropriate and feasible actions were taken."[19]

With Reagan's election, however, the warnings of *The Global 2000 Report* were simply ignored. Shortly after Reagan's 1980 victory, Republican congressmen David Stockman and Jack Kemp presented the president-elect with a detailed economic plan entitled "Avoiding a GOP Economic Dunkirk." Stockman was a boyish and wonky conservative from Michigan in his early thirties, while Kemp, in his mid-forties, had been an all-star quarterback for the San Diego Chargers and the Buffalo Bills before getting into politics as an economic conservative from western New York. Warning of economic threats facing the incoming Reagan administration, from recession to possible short-term surges in oil and food prices, Stockman and Kemp urged Reagan to defuse the "regulatory time bomb" put in place during the 1970s wave of environmental, energy, and safety legislation. "McGovernite no-growth activists" had gained control of key administrative posts under Carter and had generated a "mind-boggling outpouring of rule-makings, interpretative guidelines, and major litigation" that would create a staggering regulatory burden. Stockman and

Kemp called for a "regulatory ventilation" that would unilaterally "defer, revise or rescind" regulations that threatened to impose more than a hundred billion dollars in compliance costs. They particularly warned about the consequences of new federal standards for automobile and truck emissions, workplace noise, asbestos exposure, appliance efficiency, and industrial wastewater. The EPA had rules, Stockman said, that "would practically shut down the economy if they were put into effect."[20]

Reagan liked this antiregulatory advice so much that he appointed David Stockman director of the Office of Management and Budget to carry out the proposed agenda. The *Wall Street Journal* covered the appointment by calling Stockman a "relentless warrior against the widely held view that society is running out of resources and that government must therefore allocate them." The *Limits to Growth* viewpoint, Stockman explained, merely provided a rationale for those advocating more government economic planning. In his first month in office, Reagan followed Stockman's advice by postponing hundreds of regulations and ordering a review of potentially burdensome federal rules. Reagan also created a new cabinet-level task force on regulatory relief led by Vice President George H. W. Bush. Reagan cut funding for alternative energy development—later in his presidency, the White House solar panels that Carter had installed to great fanfare were removed and placed in storage.[21]

Reagan appointed western firebrand James Watt to spearhead the administration's attack on federal natural resource management. A tall, thin, balding western conservative, Watt had served previously under Presidents Nixon and Ford as a deputy assistant secretary of the Interior, working on water and power projects. Watt understood the federal bureaucracy and how to move it. He had a steely temperament and deep religious faith following his adult baptism and embrace of the Pentecostal evangelical church. Watt also had a clearly articulated political perspective that favored the "development of our natural resources by private enterprise." Just before joining Reagan's cabinet, Watt served as the first president of the Mountain States Legal Foundation, a nonprofit legal center funded by Joseph Coors Sr. of the Coors Brewing Company, to strengthen private property rights and contest government regulation. Coors also had founded the conservative Heritage Foundation to provide a philosophical underpinning for the anti-environmental movement. From its inception, the Heritage Foundation urged followers to "strangle the environmental movement," which Heritage named "the greatest single threat to the American economy." Watt spoke passionately against the environmental movement. In a 1978 speech in Dallas, Watt warned of a "new political force in the land—a small group of extremists who don't concern themselves with a balanced perspective or a concern about improving the quality of life for mankind—they are called environmentalists." What was the real motive for these "extreme environmentalists"? Watt suggested that their goal was to "delay and deny energy development" and to "weaken America."[22]

Newly empowered as secretary of the interior, Watt opened federal lands to development and fired departmental attorneys responsible for enforcing environmental standards. He sought to push out unsympathetic career government employees so that he could hire staff members who shared his views. Watt declared candidly that his mandate was to "undo 50 years of bad government," by which he meant the expansion of federal control and regulation of public natural resources. He sought to open public lands for energy and mineral development, and he halted the purchase of new national park lands. Watt emphasized broad and easy access to existing parks, rather than

Ronald Reagan with Secretary of the Interior James Watt at the signing of the 1982
Reclamation Reform Act. Courtesy Ronald Reagan Library.

wilderness protections that only served "elitist groups" and
"rugged young backpackers." Carl E. Bagge, president of the
National Coal Association, reportedly said of Watt's appointment, "We're deliriously happy." Representative Morris Udall,
a leading environmentalist in Congress, complained of Watt
that Reagan had picked the "most controversial, bombastic person" he could find. The *Washington Post* reprinted a joke told in
corporate suites: "How much power does it take to stop a million environmentalists? One Watt." Reagan supported Watt's
efforts and shared Watt's belief that he had to protect the American people against environmental extremism. In a September
1981 diary entry, Reagan recorded a meeting with Watt, noting,
"He's taking a lot of abuse from environmental extremists but
he's absolutely right. People are ecology too and they can[']t for-
age for food and live in caves." Reagan's selection of Colorado
state representative Anne Gorsuch to run the Environmental

Protection Agency appalled environmentalists almost as much as Watt's appointment. A former lawyer for the regional telephone company, she had been elected to the Colorado legislature in 1976, where she made her reputation as one of the "House Crazies" who sought a fundamental conservative overhaul of government. Known as the "Ice Queen" and the "Dragon Lady," Gorsuch immediately started making enemies among the EPA's career staff after her arrival at the agency. Critics—including Russell Train, the EPA's second administrator under Nixon and Ford—warned that Gorsuch's proposed personnel and budgetary cuts threatened to "destroy the agency as an effective organization."[23]

The day after Reagan's inauguration in 1981, the leaders of nine of the largest national environmental organizations met in a Washington restaurant to coordinate their response to the new administration. Watt and Gorsuch became the visible targets for campaigns that brought a surge in membership in the national organizations. The Sierra Club grew by 30 percent per year in the early 1980s, doubling in size in just a few years. A National Audubon Society fundraising appeal that directly attacked the Reagan administration yielded ten times the donations of its previous efforts. Environmental organizations filed lawsuits to compel the administration to enforce the Clean Water Act and other laws, even as the administration sought to cut the EPA's enforcement budget by 39 percent in inflation-adjusted terms.[24]

Environmental organizations became increasingly allied with the Democratic Party and with moderate Republicans, whose numbers were starting to shrink. The environmentalists flexed their political muscle effectively in their fights with Watt and Gorsuch. Both appointees had been driven from office by the end of 1983. Reagan described Gorsuch's resignation over a congressional investigation of the Superfund program as a "lynching by headline hunting congressmen." Watt had done a "fine job," Reagan thought, but the "Environmental Lynch mob" got him, too. Environmentalists celebrated. "I Survived the Ice Queen's Acid Reign," read the T-shirts distributed by alienated EPA employees after Gorsuch's resignation. With political success, however, came the recognition of a new vulnerability. Professional national environmental organizations increasingly depended on doomsday warnings to raise money to fuel their growth. Watt played such an outsized role as a bogeyman for environmentalists that his departure nearly caused a financial crisis for the Sierra Club. Michael McCloskey, executive director of the Sierra Club at the time, recalled that they "had lost the villain" that they needed to campaign against. Media coverage of Reagan's environmental policies plummeted. New member growth and charitable donations dropped sharply. The split between Republicans and environmentalists, however, continued to grow in the years following Watt's departure. Both Republicans critics of environmental regulation and environmental advocates, who leaned toward the Democratic Party, used the conflict to sharpen their public identity and enlist and motivate supporters.[25]

Paul Ehrlich considered Reagan, Gorsuch, and Watt simply uneducated on environmental issues. "I don't believe that those people are either total morons or totally evil," Ehrlich said in a 1983 interview. Therefore, he reasoned, they must simply be "profoundly ignorant." Ehrlich could not fathom the possibility that fundamentally different values or ideologies might yield different conclusions. His certainty helped make Ehrlich a more zealous advocate and steel him for political combat, but it also made it hard for him to understand his critics and persuade others. Ehrlich continued with his butterfly research, primarily in California and at the Rocky Mountain Biological Laboratory in Colorado. He wrote to the Yale ecologist G. Evelyn Hutchinson in 1983 that his research group had developed a useful new theory of butterfly mating strategies. "A small triumph, but one settles for them more and more!"[26]

Ehrlich welcomed the small scientific victories, because the Reagan years largely brought him to despair about the future. Reagan's military buildup and hawkish rhetoric particularly disturbed Ehrlich, and he turned his attention to the dangers of nuclear war and proliferation. Scientists had worked on antinuclear campaigns since the 1940s, but now they focused more on dangers to biological systems. The issue preoccupied Ehrlich. In June 1983, he wrote to Hutchinson, "I grow increasingly apprehensive as the December deadline for cruise missile/Pershing II deployment approaches with no signs of progress in the [intermediate-range nuclear forces] negotiations—and as the administration pushes forward with the destabilizing MX [missile]." Ehrlich channeled his concern about nuclear weapons into research and writing. He helped organize a 1983 conference in Cambridge, Massachusetts, on the biological threats posed by nuclear war. Ehrlich also served as lead author and chief organizer for an essay in *Science* on the "Long-Term Biological Consequences of Nuclear War." Twenty other prominent scientists, including Carl Sagan and Stephen J. Gould, signed on as coauthors. They warned that a large-scale nuclear war could cause a nuclear winter, in which soot and smoke would block sunlight and precipitously drop global temperatures. Ehrlich had raised this possibility in his 1977 *Ecoscience*. The change in solar radiation brought about by the nuclear explosions could destroy civilization's biological support systems and reduce human populations to "prehistoric levels or below."

"Extinction of the human species itself cannot be excluded." Ehrlich also coedited, along with Carl Sagan and two others, a 1984 report from the Cambridge conference, entitled *The Cold and the Dark: The World After Nuclear War.* The threat of nuclear winter was closely tied to Ehrlich's earlier apocalyptic predictions, as he feared that resource scarcity and overpopulation would spark conflicts that would lead to a global thermonuclear war.[27]

   While Ehrlich darkly contemplated the end of civilization, Julian Simon encountered a newly enthusiastic audience in Washington, where the Reagan administration brought free market advocates and critics of environmental regulation into power. Simon's controversial essay in *Science* in 1980 changed everything for him, and he relished the attention that followed. "I have hit the jackpot," he wrote in notes to himself the following year. "The world has now made it easy for me to remain undepressed. I no longer must deflect my mind from my professional difficulties in order to stay happy, but instead I can now dwell on my worldly 'success' and take pleasure from it." His article led to many invitations to write and speak. He finally had a chance to "reach a wide audience with a set of ideas that had previously fallen mostly on deaf ears, or more exactly, on no ears."[28]

   As Ehrlich and Simon argued in the pages of *Science* and *Social Science Quarterly* and agreed to their bet, Julian Simon completed the final touches on what he hoped would be his magnum opus, *The Ultimate Resource.* Published in 1981 by Princeton University Press and excerpted over three issues of the *Atlantic Monthly,* the book crystallized Simon's thinking about the relation between population and resources issues in accessible prose. In Simon's formulation, people were the "ultimate resource." "Human resourcefulness and enterprise" could meet impending shortages and solve problems indefinitely. In fact, new solutions generally would leave society "better off than before the problem arose." Simon acknowledged that his thesis was not original. Adam Smith and Friedrich Engels, as well as later writers such as Jules Verne and H. G. Wells, had "given full weight to man's imagination and creative powers" to solve population and resource problems. More recent inspirations included Simon Kuznets's national income and population research, Harold Barnett's writings on resource scarcity, and Ester Boserup's theories about agricultural innovation.[29]

Simon argued that food, land, natural resources, and energy were all becoming more abundant, not scarcer. How did he know? Rising prices and a rising ratio of price to income were the indicators of scarcity. Yet since the beginning of the Industrial Revolution, resource prices generally had fallen, particularly relative to income. Humans spent less and less of their time and income meeting essential needs for heat, light, food, and water. Simon also argued that population growth was not mechanical and automatic, as Malthus (and later Ehrlich) had suggested in long-term projections of exponential growth. Having babies was like drinking alcohol, Simon wrote. Few people were "drunkards" who drank themselves to death—most drew lessons from their experience and moderated their behavior. Similarly, Simon's research showed that people make rational choices about family size. These decisions change under different circumstances and would respond to resource scarcity and other indicators of overpopulation. While additional people certainly present a burden on society, Simon said that both new babies and immigrants "produce more than they consume." Their benefits to society far outweighed their cost. Given resource abundance and these benefits, Simon criticized misguided and often oppressive efforts to reduce population growth.[30]

Simon's critique of population control in *The Ultimate Resource* reflected his personal values and his version of utilitarian philosophy. Simon differed from Ehrlich and other population control advocates in his attitude about the worth of human life. Ehrlich, complained Simon, said that nothing would be lost if fewer people existed and that the United States, and the world, would be better off with a smaller population. In 1972, when the United States population was almost 210 million people, Ehrlich had told a reporter, "I can't think of any reason for having more than one hundred fifty million people." Simon found this dismissal of the value of sixty million Americans to be cavalier. Ehrlich's bleak descriptions of human misery further suggested to Simon that Ehrlich thought "poor people's lives are not worth living." Simon took up their cause. Writing of the impoverished beggars in India, Simon commented, "Ehrlich writes nothing about those people laughing, loving, or being tender to their children—all of which one also sees among those poor Indians." Drawing on a utilitarian perspective that aspires to the greatest good for the greatest number, Simon contended, in his own variant of the theory, that more people living rewarding lives maximized social welfare. "Because people continue to live, I believe that they value their lives. And those lives therefore have value in my scheme of things." The continued existence of poor people did not signal overpopulation, Simon wrote.[31]

The publication of *The Ultimate Resource* in 1981 established Julian Simon as a national conservative intellectual. A sympathetic reviewer in the *Washington Post* called Simon's book the "most powerful challenge to be mounted against the principles of popular environmentalism in the last 15 years."

On the public television show *Firing Line,* conservative commentator William F. Buckley declared that "Julian Simon may be the happiest thing that has happened to the planet since the discovery of the wheel." Buckley suggested that *The Ultimate Resource* would "dominate the debate" over population and resource scarcity in the 1980s.[32]

As if eager to prove Buckley right, Simon followed his book with a steady stream of articles and interviews in leading national publications, including the *Washington Post, New York Times, Wall Street Journal,* and *Los Angeles Times,* along with the *Atlantic Monthly.* Simon hammered home his argument for the economic benefits of immigration. He challenged the fear that badly needed farmland was disappearing, being paved over for highways and suburban sprawl. Instead, Simon argued that cropland was more available and less scarce than before. Julian Simon was becoming the "most visible apostle of optimism," according to the *New York Times,* with his insistence that "life on earth is getting better, not worse."[33]

Simon's rapid emergence as a public critic of environmentalism brought him to the attention of national conservative organizations that previously did not know he existed. Simon had accomplished his rise to prominence largely on the basis of his own ambition and determination and the limited opportunities and resources available to a little-known University of Illinois professor. Simon now started to benefit from financial and organizational support from foundations and conservative think tanks and to circulate in higher-level conservative circles. His appeal was no secret: Simon appeared perfectly suited to delivering and backing up an important part of Reagan's message.[34]

This institutional support became useful in 1982, when Simon teamed with idiosyncratic conservative Herman Kahn to prepare a rejoinder to *The Global 2000 Report.* Kahn was the white-bearded, rotund cofounder of the Hudson Institute, a politically conservative research center that published studies of the future. Kahn first made a name for himself with a controversial 1960 Rand Institute study, *On Thermonuclear War*, which had argued that nuclear war was both possible and winnable. As Kahn grew interested in domestic policy issues in the 1960s and 1970s, he criticized predictions of environmental catastrophe and predicted "unprecedented affluence" in the coming decades. At a 1975 conference on *The Limits to Growth* in Woodlands, Texas, Kahn declared, "Two hundred years ago mankind was almost everywhere poor, almost everywhere scarce, almost everywhere powerless before the forces of nature. Two hundred years from now mankind will be almost everywhere rich, almost everywhere plentiful, almost everywhere in control of the forces of nature." Kahn's relentlessly optimistic outlook cited the same market forces that Julian Simon emphasized—technological innovation, substitution, and discoveries of new kinds of resources. In 1980 and 1981, Kahn attacked *The Global 2000 Report* as "Globaloney 2000." At a time when the world population was 4.4 billion, Kahn declared "no reason why the world should not be able to support a population of 30 billion people." Kahn anticipated a new economic boom with abundant energy and growing international trade. He described the world as "halfway through a great transition" that would spread progress and technology "for the good of all." Kahn elaborated on these optimistic predictions at length in a full-length 1982 book, *The Coming Boom*.[35]

Simon and Kahn now undertook to reexamine the "baseless" and "gloomy assertions"of *The Global 2000 Report*. They initially sought a two-hundred-thousand-dollar consulting contract from the EPA to conduct their study. Allies within the Reagan administration, such as presidential adviser Danny J.

Boggs, shared their desire to repudiate *The Global 2000 Report*. After the news of Simon and Kahn's project leaked out, however, congressional opponents and Reagan's own White House Council on Environmental Quality blocked the contract and an official imprimatur to their report. Kahn and Simon instead turned to the conservative Heritage Foundation for financial and administrative support. Heritage's president, Edwin Feulner, feared that the Council on Environmental Quality would issue a report on global environmental trends that would "run exactly counter" to the president's point of view by "emphasizing physical limits on our progress instead of the vast possibilities for progress and improvement limited only by our energies and will." With the Heritage Foundation's support, Kahn and Simon enlisted prominent and generally conservative scholars to rebut *The Global 2000 Report*'s claims. Contributors included William Baumol, past president of the American Economics Association; John Fraser Hart, former president of the Association of American Geographers, and Aaron Wildavsky, a leading political scientist at the University of California, Berkeley. The authors were asked to contribute essays that rebutted the claims of *The Global 2000 Report*.[36]

After Kahn died unexpectedly of a stroke at age sixty-one in 1983, Simon finished editing the volume by himself. In the introduction, Simon polemically offered a direct inversion of the claims of *The Global 2000 Report*. The report had declared:

> If present trends continue, the world in 2000 will be more crowded, more polluted, less stable ecologically, and more vulnerable to disruption than the world we live in now. Serious stresses involving population, resources, and environment are clearly visible ahead. Despite greater material input, the world's people will be poorer in many ways than they are today.

Simon countered:

> If present trends continue, the world in 2000 will be less
> crowded (though more populated), less polluted, more
> stable ecologically, and less vulnerable to resource supply
> disruption than the world we live in now. . . . The world's
> people will be richer in most ways than they are today. . . .
> Life for most people on earth will be less precarious
> economically than it is now.

William Tucker, author of *Progress and Privilege: America in the
Age of Environmentalism,* described *The Global 2000 Report* as
the "wicked witch in *Sleeping Beauty*" who "laid a curse upon
the land." Simon and Kahn's *The Resourceful Earth,* in turn, had
arrived like the "good witch . . . to lift the spell." "For more than
a decade, the prophets of gloom and doom have had their way,
painting a frightening picture of the world coming apart at the
seams," wrote a journalist in a cover story in the *Chicago Tribune* sympathetic to Simon and Kahn. Their book now declared
these predictions "nonsense."[37]

A vigorous debate ensued. In the spring of 1983, Simon
and coauthors in *The Resourceful Earth* presented their findings
at the American Association for the Advancement of Science
meeting in Detroit. The session brought national attention to
the report and provoked lively discussion. "That was a good talk
in Detroit," the Harvard demographer Nathan Keyfitz, one of
Simon's critics, wrote to Simon afterward. "You are the most
interesting person with whom I have disagreed in a long time."
In December 1983, the American Economics Association invited Simon to participate on a panel at its annual meeting
in San Francisco, "Limits to Growth: What Have We Learned?"
In May 1984, Simon organized a second American Association
for the Advancement of Science session entitled "Knockdown-
Dragout on the Global Future," pitting Simon and Danny Boggs,

New Yorker cartoon highlighting the sharp contrast in views of the future in the early 1980s. © Frank Modell/New Yorker Cartoon Collection/ www.cartoonbank.com.

deputy secretary of energy, against the environmental scientists Barry Commoner and Peter Raven. "I shall attack you and your report," Commoner told Simon. The clash between starkly different ways of viewing the future became an increasingly common trope. *Science* captured it in an article about Simon and his critics. In a cartoon accompanying the article on the forthcoming *Resourceful Earth,* a man in an overstuffed chair read a book entitled "The Coming Boom." To his left sat another man, reading "The Coming Collapse." *USA Today* summarized the choices in a 1983 headline: "Future is a.) dim or b.) bright (pick one)." Simon and his opponents both used the stark contrast between their points of view to raise their public profiles. As the lesser-known challenger attacking conventional beliefs, Simon perhaps benefited the most from the attention.[38]

Simon's efforts gave political conservatives a weapon that they badly needed to support their attack on liberals and environmentalists. *The Resourceful Earth* and *The Ultimate Resource* bolstered the Reagan administration's critique of environmental regulation and of Jimmy Carter's economic record. At a speech in Texas in 1983, Vice President George H. W. Bush embraced Julian Simon's way of thinking about natural resources issues. Bush denounced *The Global 2000 Report* as a vision of stagnant economic growth and an "age of limits." He called it Carter's "economic philosophy in black and white." Bush continued, "We all have a choice to make: It is between the shrinking vision of America held by the pessimists or the expansive vision—the expansive reality—we are building right now. We are too great a nation, we are too great a people to shrink from the future." Bush's speechwriter, Joshua Gilder, wrote to thank Julian Simon for his "inspiration and research." Gilder called himself a "great fan" of Simon's work. In a 1983 commencement address at Ohio State University that Gilder also helped to write, Vice President Bush practically quoted Simon's work while discussing how the "prophets of doom" were wrong and global trends improving. Humanity was not depleting the world's resources. Instead, Bush argued, "the world's resources are becoming more plentiful all the time. The fact is that new technologies are not only allowing us to use our old resources more efficiently—they actually create new resources." Julian Simon's ideas about resource abundance also resonated with other Republican politicians, sometimes in quirky ways. Georgia congressman Newt Gingrich, who would later rise to be Speaker of the House of Representatives and run for the presidency, reached out to Simon to discuss the possibilities of using the "resources of space to counter the Limits-to-Growthers." Administrators in the Department of the Interior and the Census Bureau invited Simon to brief their staffs—or, as William Butz of the Census Bureau phrased it, "get us all riled up."[39]

As Julian Simon ventured further into the national arena, becoming a favored critic of the environmental movement and a regular presence on the op-ed pages, Rita and he decided in 1983 to leave the cornfields and quiet streets of Illinois to move to Washington, DC. Rita had been offered a position at American University as dean of the School of Justice, a department focused on the study of law and society. Julian, meanwhile, had received a grant from the Sloan Foundation to study the economic consequences of immigration, which paid half his salary for eighteen months. He became a senior fellow at the Heritage Foundation and also eventually landed a faculty appointment position teaching business administration at the University of Maryland. Having grown up in and around New York City, the Simons never imagined living all their lives in the flat Midwest. Julian also was growing a little bored with life in Urbana, where he felt "less and less excitement in the conversations that chance throws up for me." Now their kids were teenagers, soon leaving for college. Moving to Washington provided new opportunities, both personal and professional. The new location, Julian wrote to friends in 1983, "will make it more convenient to try to sell some views on public policies."[40]

Julian and Rita settled into a house in suburban Chevy Chase, Maryland. Julian enjoyed the new Washington setting. Between early spring and midautumn, Julian would sit outside on the back deck for hours with his computer. Always by his side was a pair of binoculars so that he could watch the birds visiting his feeders. At the same time, he also missed aspects of the easy life of Urbana, where he could "jump out of bed and be in the office in four minutes and then hustle out of the office

Julian, David, Daniel, Judith, and Rita Simon (left to right) at David's graduation
from high school in Urbana, Illinois, 1982. Courtesy of Naomi Kleitman.

and be on the squash court." He took to dictating correspondence and other writing into a tape recorder during his com-
mute to work.[41]

New doors continued to open for Julian Simon after he arrived in Washington. Simon used his post at the Heritage
Foundation to make inroads in the Washington policy community. In September 1983, shortly after arriving at Heritage, for
example, he invited more than twenty people to a meeting to
discuss congressional proposals to increase the government's
"foresight capability" on resources, population, and the environment. Simon opposed these congressional proposals based
on his critique of *The Global 2000 Report*. He described his
work to Burt Pines, Heritage's vice president, as "trying to put
the boots to the environmentalists' initiative for a government
'global foresight' activity." The meeting brought Simon closer

to Washington business conservatives such as Fred Smith at
the Competitive Economy Foundation (who later founded the
Competitive Enterprise Institute). Smith warned, "Conceding
any legitimacy to a government data collection role is in my
opinion extremely dangerous." Rather than more regulation,
Smith called for "free market environmentalism" led by private-
sector entities through market mechanisms and private property rights. New contacts with conservatives like Smith accelerated Simon's transformation from isolated intellectual in
Urbana, Illinois, into connected and influential Washington
commentator.[42]

From his new perch at Heritage, Simon also launched what
he called a "full-scale investigation" of the United Nations Fund
for Population Activities and the Agency for International Development. He aimed to "lay bare the patterns of funding that
wind up with" nongovernmental organizations promoting
family planning in the United States. In the fall of 1983, Simon
unsuccessfully sought credentials as a US delegate to the United
Nation's population conference to be held in Mexico City in August 1984. In a letter to the deputy secretary of state, Simon
practically begged to go, saying, "This is the first time in my
life, literally, that I have sought an appointment other than
an ordinary job." Although he did not get the credentials he
sought, Simon supported the Reagan administration's new positions on population, which rejected the idea of a population
crisis. In June, before the United Nations meeting, Simon testified before the House of Representatives Subcommittee on
Census and Population that "overpopulation" was a "myth."
Simon complained in his testimony that the US delegation did
not reflect the full diversity of views on population growth. He
warned that the "population lobby" used the issue of access to
abortion as a cover for its population control agenda. White

House staffers acknowledged that Simon's ideas influenced the administration's new global population policy. James Buckley, the chairman of the Mexico City delegation, later recalled how the Americans had been able to "pierce the Malthusian gloom" of the meeting with statistics showing the fall in birthrates in the developing world and the increase in human life expectancy, rise in caloric intake, and growth in per capita income. In the second presidential debate between Reagan and Walter Mondale that took place in October 1984, Reagan embraced Simon's position, calling the "population explosion . . . vastly exaggerated." Simon defended the administration's policy shift, arguing that Reagan was "backed by empirical scientific research."[43]

Even as Republican political leaders moved closer to Simon's point of view on population, Simon increasingly made his arguments about population into a case for free enterprise. In a 1985 essay in the *Washington Post,* for example, Simon contended that governments and development agencies continued to focus on overpopulation as the cause of international development problems in order to avoid talking about the more obvious cause of underdevelopment—dysfunctional economic and political systems. Simon rejected Carter-era government conservation. Simon supported Reagan's call for private individuals to seek profit and to create new resources. "Listening to environmentalists," Simon wrote in *USA Today* in 1984, "you'd think our air is unbreathable, our water is undrinkable, and that this country faces a crisis of major proportions. It's simply not so." Simon's narrow focus on the economics of population and resources expanded. In a 1984 essay in *Reason* magazine, Simon theorized that humanity had evolved culturally so that "our patterns of behavior . . . predispose us to deal successfully with resource scarcity." Over centuries, social rules and customs gave humanity "greater rather than less command over resources." Simon believed that humankind was "on balance a creator rather than a destroyer." His perspective contrasted strikingly with his opponents, he said, who viewed people primarily as destructive consumers of resources.[44]

Julian Simon's celebration of people's creative abilities, rather than their destructive and wasteful tendencies, encouraged his increasing focus on immigration. Paul Ehrlich had joined the immigration debate in 1979 with his book *The Golden Door*, which had sought to justify a restrictive immigration policy. Simon had countered with pro-immigration testimony to a congressionally authorized committee that same year. Immigration policy served as a natural continuation for Simon's and Ehrlich's battles about overpopulation, since both fights centered, in part, on the question of whether there were too many people. During the early 1980s, political conflict over immigration deepened. Millions of illegal immigrants crossed the nation's southern border to live and work in the United States. The *Wall Street Journal* warned against "The Latino Tide" in June 1984. "Our nation has lost control of its borders," Ronald Reagan declared the following month.[45]

Julian Simon sought to counter the stigma associated with immigration and argued instead for its economic benefits. In an op-ed in the *New York Times*, Simon attacked the idea that immigrants caused job losses. He argued instead that they helped create new jobs by expanding aggregate economic demand and creating new businesses that employed workers. In 1985, Simon published a monograph on the economic effects of immigration, which he extended into a full-length book in 1989.[46]

Simon celebrated immigration for some of the same reasons that he embraced population growth more generally. Immigrants did not threaten the US economy and society. Illegal immigrants, in particular, contributed more than they took. Illegal immigrants use few medical or welfare services because they are afraid of being found out, Simon declared in his interviews with journalists. Noting that illegal immigrants paid payroll taxes without receiving Social Security benefits or income tax refunds, Simon argued, "We rip them off unconscionably." Harold J. Barnett, the economist who had inspired Simon with his 1963 *Scarcity and Growth,* concurred that "it simply is not true that immigrants make us poorer." Simon argued further that immigrants offered a way to *strengthen* the US economy. Just as he had pushed for an auction for airline tickets, Simon suggested that the United States auction off rights to enter the country. Liberals denounced his idea, saying that it betrayed the values of the nation and the preferences given to refugees, relatives, and skilled workers. But Simon viewed immigration as a way to turbocharge the economy with new entrepreneurial citizens. His idea would later be adopted in modified form in the 1990 immigration act, which provided visas for immigrant investors.[47]

Simon's argument that immigrants benefited the American economy countered growing anti-immigrant sentiment. It also complicated liberal and conservative battle lines on the contentious subject. "Nine Myths About Immigration," an essay written by Simon and promoted by the Heritage Foundation in 1984, circulated widely on Capitol Hill. Senator Edward Kennedy, a leading liberal, embraced Simon's report and entered it into the *Congressional Record.* "We have heard that immigrants are 'welfare abusers,'" Kennedy declared, "that undocumented aliens heavily use welfare services, and that immigrants pay less than their share of taxes." But these accusations were simply "based upon fear." Kennedy quoted Simon's conclusion

that "many of the alleged costs of immigrants are simply unfounded, hollow myths." Simon welcomed Kennedy's enthusiasm and his efforts on behalf of immigrants, telling the senator that "a long ethnic memory can have benefits for all of
mankind."[48]

As Kennedy's warm Democratic embrace suggests, Simon's Heritage report and his positions drove a wedge into the
conservative coalition over the controversial topic of immigration. Some conservatives joined Simon in favoring looser immigration policies, including amnesty for some illegal immigrants, out of free-market principle or probusiness sentiment;
others feared the cultural and economic costs of immigration
and sought to expel illegal immigrants and shut the borders.
Immigration opponents complained to the Heritage Foundation that Julian Simon's "Nine Myths" was being used as "part
of a vicious campaign against the administration-supported . . .
immigration reform bill now before the House."[49]

The Reagan administration, just like the Heritage Foundation, was pulled in different directions on immigration. A 1986
draft study by Reagan's Council of Economic Advisors reflected
Simon's view that restricting immigration would hamper economic growth. But the administration's probusiness position
ultimately gave way to political pressure to reduce immigration. Simon himself opposed the administration-supported
immigration reform bill, which aimed to reduce the number of
immigrants into the United States. The 1986 Immigration Reform Act restricted future immigration levels while also providing amnesty to current illegal immigrants. Simon thought that
the immigration restrictions represented "economic ignorance
and plain racism." He called concern over the illegal status of
immigrants "mostly a red herring on the part of those who are
simply anti-Mexican." Simon acknowledged the split between

advocates of free markets and conservative anti-immigrant groups, saying that the divide illustrated why he was not a down-the-line political conservative. Simon's unorthodox position on immigration ultimately led to him shifting his affiliation from the Heritage Foundation to the Cato Institute, which was more committed to free market ideology.[50]

In addition to his enthusiasm for the free movement of labor, Simon's cherished memories of growing up in the Weequahic neighborhood of Newark clearly influenced his proimmigrant sentiments. Simon freely acknowledged that his "values and tastes favor having more immigrants." He explained, "I delight in looking at the variety of faces that I see on the subway when I visit New York, and I mark with pleasure the range of costumes and languages of the newspapers the people are reading." Accounts of immigrants moving to New York City filled him with nostalgia as he recalled the contribution that his grandparents had made "with little except their hopes and their willingness to work hard and take chances."[51]

In the years following his 1980 essay in *Science,* Simon thus established himself as a politically significant conservative thinker whose writings reshaped national debates over population, resources, and immigration. As a marker of his new prominence in the nation's capital, the *Washington Post* profiled Simon in 1984 and then again, under the title "The Heretic Becomes Respectable," in 1985. Simon was not alone in capturing the public stage as a conservative darling during the early Reagan years. Others also similarly made a name attacking environmentalists. In a prominent 1984 book, *The Apocalyptics: Cancer and the Big Lie,* for example, the journalist Edith Efron denounced environmentalists, scientists, and the media for spreading the idea that synthetic industrial chemicals were causing a cancer epidemic. Julian Simon kept pace with peers like Efron, carving out a distinctive area of expertise. Simon's demanding work ethic generated a steady stream of publications elaborating on the themes of *The Ultimate Resource*. With support from the Heritage Foundation, Simon wrote dozens of op-ed articles for leading national publications. He also landed prime-time interviews on television shows such as PBS's *MacNeil-Lehrer NewsHour* and William F. Buckley's *Firing Line*. Overpopulation debates proved a little too "abstract" and "philosophical" for more general interest shows, but Simon also tried to get on *Late Night with David Letterman*, *The Merv Griffin Show*, and *The Phil Donahue Show*. Heritage staff members helped organize attention-grabbing publicity for *The Resourceful Earth* and for Simon's immigration reports. After his years struggling to get attention, Simon marveled at Heritage's ability to do "repeatable magic" in drawing attention to his work. Simon testified before Congress and got to know conservative politicians like Jack Kemp.[52]

Simon's media appearances and popular writings invariably provoked outrage. Some critics attacked his data and conclusions. Others mocked and dismissed Simon as a "mail order master," because of his successful 1965 book on the mail-order business, which McGraw-Hill reprinted five times. Even scholars sympathetic to Simon's views criticized his controversial tone and his ideological affiliation with conservative organizations such as the Heritage Foundation and Manhattan Institute. The geographer Gilbert White, for instance, complained that the draft introduction to *The Resourceful Earth* was "needlessly contentious," undermining its credibility.[53]

Simon defended his provocative approach as necessary for an outsider who had struggled for many years to be heard. Getting to the point of publishing *The Ultimate Resource* had been a "long and difficult time for me," he wrote one colleague. To Albert Rees, president of the Alfred P. Sloan Foundation, Simon explained, "If I had not stated my arguments provocatively, starting with an article in *Science* in 1980, I think that I would still be quite on the outside, having to struggle to round up my children and a few neighbors to hear what I had to say on my chosen subject." To the demographer Samuel Preston, Simon argued, "It would be nice to have the luxury of being above the fray, striking a graceful stance and having all one's dignity. But people with minority views don't have such a luxury." In a letter to friends in 1987, Simon referred to "the pain and frustration and failure that I feel almost constantly in connection with the demographic and economic research and writing." Attacks on his scholarship and his character—and even more his recurring tendency to feel disrespected and ignored—wounded Simon, feeding the oppositional attitude that he had nurtured since childhood.[54]

Despite his recurring feelings of failure, Simon had changed the political debate in Washington through persistence and provocation. The clout of the Heritage Foundation had helped him find an audience in the newly receptive political climate. In 1986, the National Research Council, the research affiliate of the National Academy of Sciences, demonstrated just how far the population-resources debate had shifted in Simon's direction when it published *Population Growth and Economic Development*. The National Research Council had previously examined population issues in 1971, issuing a sharp warning about how population growth threatened to slow per capita income growth, deepen economic inequalities, and otherwise undermine societal welfare. The new 1986 report sought a middle road, rejecting both the "most alarmist" and the "most complacent" views regarding the economic effects of population growth. According to the economists and other social scientists involved in preparing the report, Simon's constant refrain had prompted the new literature review. *The Limits to Growth* and *The Global 2000 Report* had suggested "much to fear about population growth." By contrast, the 1986 report noted, despite rapid population growth, developing countries had achieved unprecedented per capita income, life expectancy, and levels of literacy over the previous quarter century. No clear statistical association existed between population growth rates and per capita income growth. Human behavior and institutions mediated between population and the economy.[55]

The most important scientific body in the country was saying that population growth did not present a major obstacle to economic development. The 1986 report respectfully referenced Julian Simon's work numerous times and accepted the overall gist of his arguments. The report found that concerns about resource exhaustion had "often been exaggerated." "The scarcity of exhaustible resources is at most a minor constraint on economic growth in the near to intermediate term." Price increases would spur conservation, improved extraction, and substitution. The report declared that "exhaustible resource depletion does not seem likely to constrain world economic growth in the foreseeable future." Following Ester Boserup's arguments about agricultural innovation, the report pointed out that technological advances came about through scarcity, which stimulated "a search for economizing strategies."[56] One reviewer wrote that the report was "one long subterranean roar, rumbling out 'Malthus was wrong.'" The report did not mention Malthus's name, instead trying to "slip the old man into the ground unnoticed."[57]

The National Research Council report made mainstream Julian Simon's thinking about the relation between economic and demographic change. Simon welcomed the report, saying that it "bravely wrests itself from many unsound propositions published widely in the past." Yet he was not satisfied. The report's conclusions felt to Simon like "being charged with first-degree murder when one is innocent, and then having the court reduce the sentence to manslaughter." Was he supposed to be grateful? The National Academy of Sciences had "backed away from what it now regards as the crazies . . . but has still left an unsound impression." The authors continued to insist that economic development would be faster if fertility were lower. Simon said that this incorrect view underlay "misguided and dangerous policies" of the United States. The report's soft tone allowed the World Bank, Agency for International Development, and other entities to continue their population control efforts as before. Simon was furious about the press release for the report, which attributed famine and starvation in Ethiopia to "very badly functioning markets combined with rapid population growth." That description, he said, differed greatly from an alternate account that the "food shortages were caused by dictatorial governments which beggared farmers by appropriating their land and heavily taxing their output, together with denying them the right to move freely to wherever they wished to work and live." Where the press release cited "market failure," Simon saw government tyranny in Ethiopia.[58]

Where Simon grudgingly celebrated a partial victory, Paul Ehrlich was apoplectic about the "incompetent" population report. The National Research Council study asserted that the "most important resources are not natural, but artificial," including social and economic infrastructure. Ehrlich denounced this mentality. He said the attitude ignored the degradation of land and water resources, the importance of biodiversity, and the ability of the environment to absorb pollution. Although the study focused explicitly on the economic impact of population growth, Ehrlich complained that the review committee had included no ecologists, evolutionists, or earth scientists.[59]

Many other scientists shared Ehrlich's disparaging views of the 1986 report. G. Evelyn Hutchinson asked Ehrlich to "use my name in any way that seems useful to combat this idiocy." Ansley Coale, coauthor of an influential 1956 study that had described population growth as an impediment to economic growth in developing countries, sharply criticized this new approach. He said that the report relied too heavily on assumptions of neoclassical economics that markets would address demographic challenges and did not show that fertility reduction was undesirable in low-income countries. Environmental economist Herman Daly also criticized the report as trapped by the "mental straitjacket" of neoclassical economics. The report simply ignored the constraints of long-term carrying capacity, Daly said. Daly rejected the idea that capital could replace natural resources—a "notion that cannot withstand even a moment's reflection." Daly spoke favorably of the Chinese population policy, which rejected market solutions in favor of "stringent population controls." Daly confessed his "astonishment" that a committee of the National Academy of Sciences would favor conservatives like Julian Simon and Herman Kahn over biologists such as Paul Ehrlich and Garrett Hardin.[60]

For Ehrlich, the 1986 report illustrated the dismally low status of population biology, since no population biologists participated in the writing of the report. Ehrlich argued that one reason for population biology's low status was that its results pointed toward "constraints and limits on the human enterprise." Population biologists were "seldom the bearers of good news." They told hard truths that economists, developers, politicians, and chemists needed to hear. Molecular biology might cure cancer and extend the life expectancy of Americans by "a few years at most." But failing to heed the lessons of population biologists could "easily reduce American life expectancy by 30 years or more." Unlikely events, such as nuclear war or rapid climate change, posed "nearly infinite" risks that a conservative society would avoid. Unless checked, the increasing scale of human activities, Ehrlich declared, "will lead inexorably to lower standards of living, less healthy lives, and quite likely the collapse of civilization." Ehrlich urged ecologists and population biologists to represent their interests in Washington more effectively and to train graduate students in "scientific politics."[61]

To promote biological thinking about social problems, Ehrlich worked with scientific colleagues in 1987 and 1988 to start a new group called the Club of Earth. The Club of Earth, which included a small number of other prominent biologists such as G. Evelyn Hutchinson and Edward O. Wilson, took its name from the Club of Rome, which had published *The Limits to Growth* in the early 1970s. Ehrlich thought that the group could provide "an authoritative counter-balance" to the "idiocy you get from economists and politicians." In September 1988, the Club of Earth issued a public statement warning about the problems of human overpopulation. Calling the planet "already overpopulated," the statement described population growth as second only to the threat of nuclear war as a problem facing humanity. "The population explosion will come to an end one way or another and likely within the lifetimes of most people today. The only question remaining is whether we will halt it ourselves by limiting births or whether it will be halted for us by some combination of ecological collapse, famine, plague, and thermonuclear warfare."[62]

Ehrlich's new Club of Earth failed to garner significant attention and did not continue. Ehrlich also proposed creating a

Paul and Anne Ehrlich in London, 1986. Courtesy of Sally Kellock.

new Population Biology Institute to press for funding for population biology and to help translate biological lessons into public policy. His organizational efforts revealed his increasing frustration with the public discussion of environmental issues. Population and ecological studies also were becoming marginalized within biology, as molecular and cellular approaches became dominant. Ehrlich's attacks on his critics became even more caustic. "People who don't understand why I emphasize population are the ones who have to take their shoes off to count to 20," Ehrlich liked to say. He particularly lamented the attention that the media paid to Julian Simon. Ehrlich refused to mention Simon by name, referring to him disparagingly as a "specialist in mail order marketing." More generally, Ehrlich denounced the "narrow training of economists" that made most "utterly unequipped to understand the ecological underpinnings of economic systems."[63]

Despite his accomplishments, Julian Simon felt frustrated, too. He also had tried to start a new advocacy group to promote his point of view. After the 1984 Mexico City population conference, which he had been disappointed not to attend, Simon proposed a new organization to "celebrate human life." The new entity, alternately called Pro People or Committee on Population and Economy, would counter the population establishment and show that there was no "consensus" on the need for population control. Sensing a "shift in the wind" after two decades of media despair about overpopulation, Simon wanted to remind people that "children are the heart of progress" and the "measure of all things in the Jewish-Christian-Islamic-Western tradition." "An additional human being tends to benefit rather than harm others economically." Limits to human progress were receding, and population growth, on average, increased the standard of living rather than reducing it. Pro People would oppose legislative efforts to discourage parents from having children and to stabilize the United States population. Most importantly, Pro People would provide the media with "an organizational address" that could provide contrasting viewpoints on population issues. In the absence of such an organization, Simon complained, "the anti-natalists frequently characterize those who do not agree with them as a tiny and bizarre fringe group." Pro People would "combat the anti-population and anti-growth ideas" of groups such as Zero Population Growth and the Global Tomorrow Coalition, which was promoting *The Global 2000 Report*. Pro People, like Ehrlich's Club of Earth, went nowhere organizationally. Simon continued to lack strong organizational support for his views. The Heritage Foundation proved too tactically focused on the short-term for Simon, whose independent and scholarly attitude did not mesh well with the action-oriented, policy-focused think tank.[64]

Simon's frustration came from his feeling that he had won the intellectual argument but failed to change policy behavior or break up the population establishment. The 1986 National Research Council report provided a new scientific synthesis that largely repudiated Ehrlich's views on population growth. Simon described the shift on population issues as "the unreported revolution." Population control advocates had the same prominent media platform for their views. In May 1989, Ehrlich presented three five-minute segments on NBC's *Today* show, reaching an audience of millions. The television segments, Simon complained, showed "nary a whiff of the 'balance' that journalists pride themselves on." What could be done about this? Simon had little hope. "Efforts to change the beliefs of the public and the assertions of journalists," he wrote, "are likely to be a waste of time."[65]

Powerful governmental institutions also continued to urge population control. The 1986 National Research Council report had not left much of a mark on international population policy. In a talk at the World Bank in 1988, Simon asked, "How can the World Bank, the UNFPA, AID, Planned Parenthood, and the population establishment go on repeating the same old scary statements?" Simon took his contrarian views right into hostile territory. The bank's president, Barber Conable, had recently called population control "imperative" for developing countries. Simon denounced that point of view. People like Conable, Simon said, showed "blatant intellectual dishonesty" or turned "a blind eye to the scientific evidence." Simon now broadened his attack to characterize population control advocates "as warriors against human life, even as enemies of humanity," because they aimed to prevent people from being born. "Stupidity in high places—including the lofty places here in this World Bank Building—has cost the lives of tens of millions or even hundreds of millions of human beings in the last decade or so, far more human lives than were lost in World War II." Simon blamed "simple racism," a "corrupt" relationship between researchers and policy-makers, and the desire to avoid divisive political and economic reforms for the continued embrace of population control. The "world's problem," Simon concluded, was "not too many people, but lack of political and economic freedom."[66]

As the 1980s came to a close, the decade's events validated Reagan's optimism in many respects. While the American economy had its ups and downs and ended the decade dipping into recession in 1990, economic growth generally was strong and unemployment relatively low. When the Berlin Wall fell in 1989, it marked the end of decades of Soviet domination of

Eastern Europe. Perhaps most important, no global food or energy shortages occurred or even appeared on the horizon. At
the same time, there was much to criticize in Reagan's soaring
defense spending, the expanding federal deficit, and growing
economic inequality. Reagan's broad antagonism toward 1970s
environmental regulation also was out of step. Public commitment to laws protecting clean air and water and public health
proved fiercer and more resilient than Reagan anticipated. By
provoking a forceful and organized backlash, his administration's stark anti-regulatory rhetoric in the end may have made
it more difficult to achieve meaningful conservative reforms.
Reagan failed to fundamentally change the course of environmental law. He ended up appointing more moderate environmental administrators after 1983 and, heeding the warnings of
scientists, even forcefully embraced the 1987 Montreal Protocol
for the protection of the ozone layer. Substantive policy changes
in the environmental arena in the 1980s thus were relatively
modest and ultimately forced toward the center.[67]

Yet the ideological battles of the Reagan years scarred the
nation. The growing environmental divide between the two
political parties and within the nation as a whole mirrored the
gulf between Ehrlich and Simon. Each man believed that he
was losing his life's intellectual struggle in the political arena.
Each had ventured far into the political fray, straying well beyond the narrow confines of academia. Simon's move to Washington had brought him close to powerful people. He had helped
overturn prevailing views on population growth as well as immigration. He had contributed to the rejection of the Carter
administration's environmental and economic viewpoints, em-
bodied in *The Global 2000 Report*. Ehrlich, in turn, had continued to stoke the fires of the environmental movement. He had
denounced James Watt for his environmental sins and bashed

Reagan for his high-stakes nuclear brinksmanship. Ehrlich warned of the mass extinction of species and the collapse of human civilization in a nuclear winter. Could the two men, and the competing camps that they represented, find a way to reconcile their views, or would they descend further into bitterness and recrimination, talking past each other in a fury of argument and counterargument?

CHAPTER FIVE

Polarizing Politics

O ne day in October 1990, Julian Simon picked up his mail at his house in suburban Chevy Chase, Maryland. In a small envelope sent from Palo Alto, California, Simon found a sheet of metal prices along with a check from Paul Ehrlich for $576.07. There was no note.

Simon had prevailed in their bet, by every measure. Despite a record increase in the world population from 4.5 to 5.3 billion people, the prices of the five minerals—chromium, copper, nickel, tin, and tungsten—had fallen by an average of around 50 percent. Twenty years after he had publicly attacked Paul Ehrlich in his Earth Day remarks in the University of Illinois auditorium, Simon finally had a tangible, definitive victory that he and his supporters could crow about. It was only several hundred dollars, but its value in bragging rights was priceless.

Ehrlich and Simon's bet quickly became a symbolic weapon in a long-running ideological battle between conservative critics of regulation and environmentalists. For conservatives and libertarians, Simon's victory proved that environmentalists were fools. The journalist John Tierney brought national attention to the bet in 1990 when he wrote about it in the *New York Times Magazine*. Tierney was a personal fan of Julian Simon rather than a neutral observer. He would later credit Simon with the "most fruitful telephone call of my reporting career." The call came in 1985 when Tierney interviewed Simon for a story about population growth in Kenya. Tierney thought he would hear that Kenya's overpopulation had created a hopeless situation. Instead, Simon said, "Isn't it wonderful that so many people can be alive in that country today?" Simon's radical alternative perspective on population had brought Tierney up short—and converted him into a population optimist. Ever since that call, Tierney later said, the "world has seemed a much cheerier place." Tierney and Simon became friendly, occasionally playing squash in Washington. Tierney saw Julian Simon as "a mentor and friend" who had given him a "whole different way to look at" environmental and population issues.[1]

In his 1990 telling of the bet in the *New York Times Magazine*, Tierney portrayed Ehrlich as a fraud, a MacArthur Fellowship–winning jet-setter whose predictions—about almost everything—were spectacularly wrong. Tierney located Ehrlich at the end of a trail of failed prophecy that started with Thomas Malthus and continued through the nineteenth-century British economist William Stanley Jevons to the twentieth-century American natural scientists Fairfield Osborn and William Vogt. Typical arguments over these predictions involved disagreements over data or far-off events that would not take place for decades or even centuries. Ehrlich's bet with Simon—with its clear winner based on specific commodity prices—provided a rare chance to "chart and test the global future," Tierney wrote.

Ehrlich, of course, disagreed. He insisted to Tierney that the bet did not prove anything. The only mistake he had made was in the scope of the ten-year timeline. The prices of those

Julian Simon with objects made from the metals in the Ehrlich-Simon bet, 1990.

Dan Borris/Edge Reps.

five commodities would rise. It would just take a little longer. And then it would become clear that "Julian Simon is like the guy who jumps off the Empire State Building and says how great things are going so far as he passes the 10th floor." Ehrlich warned of the new problems of ozone depletion and global warming, and the planet's ability to handle human impacts. But Tierney took Simon's side about environmental trends and the future. He cast Simon as an underappreciated, unassuming, and generally good-natured truth teller who had steadily gained the upper hand in the rivalry because of the strength of his ideas.[2]

After Tierney brought national attention to the bet, Ehrlich's public defeat provided regular fodder for conservative critics of environmental regulation during the 1990s. The Ehrlich-Simon bet confirmed "cornucopian claims that the supply of resources is becoming more abundant, not more scarce," wrote libertarian economist Ron Bailey in his 1993 book *Eco-Scam: The False Prophets of Ecological Apocalypse*. Bailey argued elsewhere that the problem with the environmental movement was a "failure of theory"—an overreliance on flawed models that yielded exaggerated predictions of doom. In *God Wants You to Be Rich*, financial advice author Paul Zane Pilzer celebrated Simon as a "true genius" and denounced Ehrlich as a "charlatan." Pilzer's book shared his vision of a world without physical limits that offered endless opportunities for wealth. The "Malthusian era can be said to have ended officially" when Simon won the bet in 1990, Peter Huber declared in *Hard Green: Saving the Environment from the Environmentalists—A Conservative Manifesto*.[3]

The details of mineral prices mattered less to conservative commentators than the simple indisputable fact of Simon's victory over Ehrlich. Yet what actually happened with the five

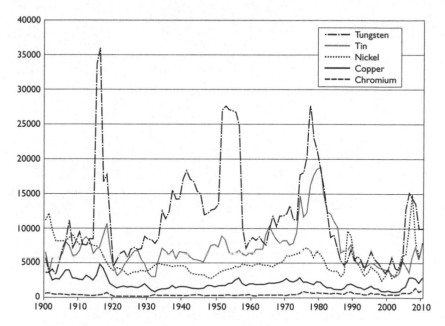

Prices of individual metals from the Ehrlich-Simon bet, 1900–2010 (1980 dollars).
Price data from Thomas D. Kelly and Grecia R. Matos et al., "Historical Statistics for
Mineral and Material Commodities in the United States," US Geological Survey,
Data Series 140, Version 2011.

minerals is still important, even if prosaic. In part, the price
changes simply reflected the ups and downs of the global economy, especially oil prices, and on this score Julian Simon had
been very lucky. When the global economy fell into recession in
the early 1980s, economic activity slowed and demand for minerals dropped. After several years of reduced demand and oversupply, mineral prices rose again in 1986 and 1987, although
not to their 1980 heights. Then the stock market crashed and
economic growth slowed. Mineral prices fell again, ending the
1980s significantly lower than where they started. Macroeco-
nomic cycles, far more than population growth, had governed

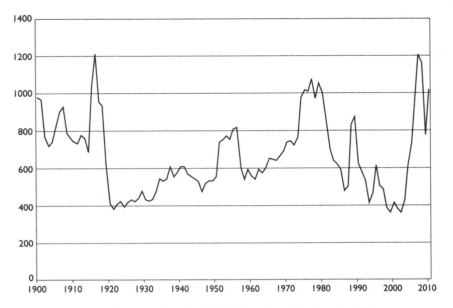

Index of the five metals from the Ehrlich-Simon Bet, with 1980 price of $1,000
($200 of each metal). Price data from Thomas D. Kelly and Grecia R. Matos et al.,
"Historical Statistics for Mineral and Material Commodities in the United States,"
US Geological Survey, Data Series 140, Version 2011.

the rise and fall of commodity prices over the course of the decade.

The "everyday market forces" that Simon, Robert Solow, and other economists had cited in response to *The Limits to Growth*—technological innovation and substitution, price-driven competition, and new sources of supply—also helped lower commodity prices. Each mineral had a slightly different story. Tin was used in the late 1970s and 1980s largely to coat the interiors of steel containers, as well as for soldering electronic components. International tin prices depended heavily on agreements that tin-producing countries had made to limit production and sustain higher prices. During the late 1970s, tin prices rose sharply. Many observers anticipated shortages. Malaysian producers sought to corner the tin market to force prices even higher in 1981 and 1982. New high-quality tin deposits discovered in the Brazilian Amazon, however, undermined this market-cornering effort. By the end of the 1980s, Brazil, a previously marginal producer, produced roughly one quarter of the world's tin supply. At the higher prices of the late 1970s, demand for tin also started to fall. Manufacturers substituted aluminum and plastic for tin in packaging. Trying in vain to stabilize prices, the International Tin Agreement held a quarter of the world's annual tin production off the market. The International Tin Council soon ran out of money, however, and the agreement collapsed. Tin prices went into "free fall," dropping more than 50 percent from $5.50 per pound in October 1985 to $2.50 per pound in March 1986. Overall, between 1980 and 1990, the gyrations of the tin market supported the arguments of Simon and the economists. New sources of supply, product substitution, and, above all, the breaking of the tin cartel had a far greater impact than population growth on tin prices, ultimately driving them down almost 75 percent.[4]

As with tin, a "copper fever" in the commodities markets similarly sent the futures market and copper stocks surging in the late 1970s. Copper was used widely for plumbing and wiring in construction, appliances, and automobile manufacturing. Many enthusiastic investors shared Ehrlich's faith that copper's rising price reflected its finite supply. Yet the price increase of the late 1970s resulted partly from labor strikes at Chilean copper mines and political disruption in copper-producing Zaire and Zambia. In the summer of 1980, prices dropped again because of recession and declining demand. New copper producers also entered the market. A worldwide surge in copper production coincided with the 1981 recession, depressing copper prices for nearly five years. Copper also saw new competition from fiber-optic cables and satellites, which telephone companies started using for their new communications infrastructure. Aluminum and plastic replaced copper in other uses. By 1985, a Chilean study of the world copper market described the industry's problem as one of "how to create demand." Devastated US producers shut down one-third of their copper production. Investment stalled, bringing another upswing in prices near the end of the decade. Yet copper still ended the 1980s nearly 20 percent cheaper than it started. As with tin, everyday market dynamics had outweighed population growth as a determinant of copper prices.[5]

As most economists knew intuitively, and as Ehrlich learned painfully, volatile commodity prices served as a poor proxy for the impact of population growth, certainly over the course of just a decade. Subject to so many competing forces, commodity prices frequently cycle from boom to bust and from scarcity to overabundance. Ehrlich's bet had been foolhardy—as even he later admitted. The bet also suggested that Ehrlich and his colleagues only tenuously understood economics and commodity markets.[6]

At the same time, Simon's victory was not guaranteed. He would not have won *any* ten-year bet about commodity prices. Indeed, when economists later ran simulations for every ten-year period between 1900 and 2008, they found that Ehrlich would have won the bet 63 percent of the time. That did not mean, however, that a longer bet over the course of the century necessarily would have gone to Ehrlich. Rather than a steady increase in prices over the century, the 1900–2008 data showed a precipitous crash in commodity prices after World War I and then a century-long irregular return to World War I levels. The simulations showed Ehrlich prevailing in 63 percent of the ten-year bets largely because prices had plummeted so low in the post–World War I crash.[7]

And yet the symbolism of the bet itself proved simple. The stark clash of perspectives that the bet represented suited the divisive environmental politics of the early 1990s. Regulations to protect endangered species, policies to slow global warming, and efforts to protect national forests and rangelands now sharply split Democrats and Republicans. Whereas in the 1970s, major environmental legislation had passed with bipartisan support, by the early 1990s, where a politician stood on environmental policies served as a litmus test of ideology and political affiliation.

In part, this reflected the increasing ideological polarization of the parties, which gradually began to sift themselves into homogenous regional and ideological blocs. The South and the Plains states became increasingly Republican, whereas coastal state congressional representation moved toward the Democrats. Along with abortion rights, gun control, and race relations, environmental policies mobilized both liberals and conservatives for all the reasons that they divided Ehrlich and Simon. Americans did not seem to vote explicitly on environmental questions in the same way that they mobilized around other controversial issues. A candidate's position on the environment instead reflected a broader set of basic assumptions about the future, about the role of government in the economy, and about the relationship of humans to the planet. Congressional voting records on environmental legislation since the 1970s reveal the steadily growing separation between the two parties after 1980.

Republicans elected during the Reagan years increasingly questioned the science and economics of environmental regulation.[8]

The presidency of George H. W. Bush initially promised to break this hardening Republican opposition to environmental policies. Bush represented a pre-Reagan kind of Republican. He was born into a blue-blood northeast Republican family, the son of Prescott Bush, a Yale-educated businessman and senator from Connecticut. Prescott Bush had been active in the American Birth Control League and Planned Parenthood in the 1940s. As a congressman from Texas in the early 1970s, George Bush followed his father's path on population issues and cosponsored federal legislation to provide birth control and family planning services through local clinics. He also chaired a 1970 Republican task force on resources and population, and he urged the stabilization of the US population. In the 1980 campaign, George Bush had dismissed Reagan's hard-right conservatism as "voodoo economics." Bush served Reagan loyally as vice president for two terms, including chairing a task force that sought to reduce the burden of federal regulations. But as he launched his own 1988 campaign for the presidency, Bush broke with Reagan's environmental legacy. He declared his intention to be the "environmental president." Bush toured Boston Harbor by boat to denounce his Democratic opponent, Massachusetts governor Michael Dukakis, for not doing more to combat water pollution.[9]

Once in office, George Bush appointed moderate Republicans with strong environmental credentials to leadership roles. William K. Reilly, the new administrator of the Environmental Protection Agency, contrasted strikingly with James Watt and Anne Gorsuch, the western firebrands Reagan had embraced in 1981. After growing up largely in the East, Reilly went to college at Yale and got a law degree from Harvard. Reilly served briefly as a military intelligence officer in Europe in the late 1960s and then returned to the United States to go back to school, completing a master's degree in urban planning at Columbia. He spent part of his time at Columbia in 1968 listening to the antiwar student protestors who shut the campus down. Reilly joined Nixon's White House Council on Environmental Quality in 1970. He was one of the agency's first staff members and helped draft early federal environmental regulations. After his early stint in government, Reilly served as president of two centrist environmental organizations, the Conservation Foundation and the World Wildlife Fund, between 1973 and 1989. Where Reagan's appointees Watt and Gorsuch saw dismantling the federal regulatory state as their political mission, Reilly had helped to create the regulatory agencies. Reilly rejected the idea that economic growth inherently conflicted with environmental protection. He actively identified himself as an environmentalist, and he sought to break the contentious stalemate that had paralyzed environmental policy-making during the Reagan years.[10]

By appointing Reilly to run the EPA and placing moderates in other key environmental positions, Bush thus sought to reconnect publicly with Nixon-era Republican environmentalism. At the same time, other Bush appointees leaned in the opposite direction, including John Sununu, Bush's chief of staff and the former governor of New Hampshire. Sununu stressed the economic costs of regulation. He had a doctorate from MIT in mechanical engineering, and he viewed climate models skeptically, questioning whether costly proposals to cut carbon dioxide emissions made sense or were being driven by "emotions." Advisers like Sununu believed that environmentalists exaggerated the risks of pollution and environmental degradation. Navigating between these two camps, Bush's environmental policies

differed significantly from litigation-oriented strategies favored
by many Democrats and environmental advocates. Bush argued, "You can get further by seeking people's help than suing
them." At a 1991 ceremony to tout an agreement to protect air
quality in the Grand Canyon area, Bush held out hope for "civil-
ity and cooperation." "For too many years, Americans have divided into feuding camps, people sparring over causes, special
interests battling it out against special interests." Bush urged
greater use of less costly informal negotiation techniques instead of lawsuits.[11]

Bush linked environmental protection with free market solutions rather than government-directed regulation. "Free markets and economic growth provide the firmest foundations for
effective environmental stewardship," Bush announced in 1991.
Bush's signature environmental achievement used markets to
improve air quality. The 1990 Clean Air Act Amendments relied on an innovative emissions trading program to reduce the
power plant pollution that caused acid rain in the Northeast.
Rather than mandate specific improvements in technology, the
emissions trading program encouraged electric utilities to reduce their emissions in the least expensive way. Each utility received allowances to emit sulfur dioxide and could either use
those allowances for their own emissions or sell them to another utility. Pollution from sulfur dioxide and nitrogen oxides
dropped dramatically as a result of the law. Of course, this struc-
turing of markets and incentives by government was not "free,"
in the sense of separate from government policy, but it was
more flexible and responsive to the marketplace. The new system encouraged less costly reductions in pollution and, by putting a price on emissions, gave firms an incentive to produce
less pollution.[12]

President Bush's initial moderate embrace of economically

efficient environmental regulation, however, did not represent the proximate future of either the Republican Party or the American public discourse on environmental problems. At the national level, at least, the battle lines defined by earlier public conflicts over population growth and resource scarcity carried over to new struggles over climate, even though the issues differed. Many scientists and policymakers, for example, tried during the early 1990s to get beyond a rhetorical framework of looming disaster to a more pragmatic discussion of the environment and the economy. Scientists writing about the potential impacts of climate change formulated their guidance to policymakers in more temperate language and with more sophisticated models than their predecessors had in the 1960s and 1970s. Rather than predict specific future catastrophes, scientists analyzing impacts for the Intergovernmental Panel on Climate Change (IPCC) wrote in conditional terms about what "could," "may," or would "likely" occur. The first report, published in 1990 with contributions from hundreds of scientists around the world, emphasized uncertain ecological relations and financial costs, and called for extensive additional research to ascertain the likely impact of climate change on ecosystems and human societies. The IPCC's discussion of scientific uncertainty and risk partly reflected a maturing of the environmental field and a more measured approach than Ehrlich's strident public linking of science, policy, and advocacy. The IPCC's conditional tones also reflected political constraints, imposed by governmental representatives who negotiated and compromised with scientists on the reports' final language. Although the IPCC's handling of uncertainty and risk improved on earlier apocalyptic warnings, the new approach failed to establish a moderate middle ground for sober discussion of climate change, at least in the public realm. Skeptics about climate science used

the IPCC reports' open discussion of uncertainty to argue that
climate science remained unsettled and governmental action
could be deferred. Lobbying and advertising efforts by fossil
fuel companies and other opponents of climate change and energy legislation amplified these skeptical voices. Efforts to address climate change in the US Congress went nowhere.[13]

In the House of Representatives, conservative Republicans
increasingly attacked environmentalists for being antigrowth
and overly pessimistic. These politicians sounded a lot like Julian Simon. In the lead-up to the 1992 United Nations Conference on Environment and Development in Rio de Janeiro, infor-
mally known as the "Earth Summit," Republican congressmen
John Doolittle from California and Tom DeLay from Texas took
to the House floor to ridicule environmental leaders. Environmentalists constructed "scare stories of looming man-made
catastrophe," Doolittle said. He attacked the new fears about
global climate change. "Outlandish predictions" about melting
icecaps and flooding coastlines made for "good movie material,"
Doolittle said, but rested on flimsy science.[14]

Doolittle singled out Paul Ehrlich and his "laughable" predictions for ridicule. "Now, Paul Ehrlich," Doolittle said, "I
cannot resist this because we have all listened to Paul Ehrlich's
foolish predictions for several decades now." Tom DeLay, the
deputy whip for the Republicans in the House of Representatives, eagerly joined Doolittle's mockery of Ehrlich. DeLay
started his career in the pest control business and had fought
with the Environmental Protection Agency—the "Gestapo of
government"—over the agency's ban on a pesticide to fight fire
ants. Nicknamed "the Exterminator," DeLay had a long history
of opposing environmental regulations. Ehrlich, DeLay now
complained, is "constantly paraded all around this country, if
not the world, as the foremost authority on the environment

and apocalypse now." Ehrlich's call for "deindustrialization,"
DeLay said, is what would happen if the world adopted a carbon tax in Rio to combat climate change. The resulting rise in
energy prices would destroy "the lifestyle of the United States
as we know it." "These people," DeLay said of Ehrlich and other
environmentalists, "want us to lower our standard of living."
Doolittle concluded with the hope that their remarks would
serve as an antidote when "the ecofreaks and the promoters
of extreme environmentalism get down [to Rio] and clamor or
march for new taxes and severe laws to be implemented."[15]

Doolittle, DeLay, and other Republican politicians successfully discouraged the Bush administration from signing onto
specific emissions targets and timetables in Rio in 1992. Bush,
though far more moderate than Reagan on environmental issues, hardened his position, aligning more clearly with the environmental critics in his party. The fall of the Soviet Union,
and the environmental devastation left behind in the former
communist states of Eastern Europe, fed Bush's celebration of
capitalism's compatibility with a healthier environment. In the
former communist countries, "the poisoned bodies of children
now pay for the sins of fallen dictators," Bush declared in his
speech in Rio. Bush looked back dismissively at the environmental predictions of the early 1970s. "Twenty years ago, some
spoke of limits to growth. Today we realize that growth is the
engine of change and the friend of the environment." As the
1992 election approached, Bush mocked the environmental beliefs of vice presidential candidate Al Gore, calling Gore "Ozone
Man." Bush declared that Gore was "so far out in the environmental extreme we'll be up to our necks in owls and outta work
for every American. He is way out, far out, man."[16]

After Bill Clinton's election in 1992, congressional Republicans united in opposition to the administration's environmental

policies. When congressional Democrats proposed an energy tax in 1993 to help balance the budget and fight climate change, Republicans blocked the effort. They then successfully used the energy vote as a wedge issue in the 1994 midterm elections when Republicans seized control of the House and Senate. Tom DeLay, who helped orchestrate the Republican takeover of the House, became House Majority Whip. DeLay now worked closely with other Republican leaders to roll back federal environmental legislation dating to the 1970s.

Environmental issues became a key dividing line in the battle between a conservative Republican leadership in the House of Representatives and Bill Clinton's Democratic White House. House Speaker Newt Gingrich, DeLay, and their colleagues attached environmental "riders" to budget appropriations bills to try to force the Clinton administration to reverse long-standing environmental policies. With the 1995 Timber Salvage Rider, for example, the Republican House leadership sought to suspend the Endangered Species Act and National Forest Management Act and to require the federal government to harvest timber on public lands. The timber provision was attached to a funding bill related to the Oklahoma City bombing, making it politically difficult for President Clinton to veto. While the Clinton administration successfully fought off many of these efforts during the 1990s, and even seemed to gain political popularity by portraying the Republicans as extremists, the battle over the appropriations riders showed the yawning gap between left and right over environmental policy. Republicans claimed that Democrats favored preservation over jobs and economic growth. The Democrats lambasted Republicans for serving corporate patrons at the expense of public health and the environment. The liberal cartoonist Garry Trudeau portrayed Newt Gingrich in *Doonesbury* in 1996 as a bomb with a lit fuse eagerly auctioning off "despoiling rights" to the Arctic Wildlife Refuge and public lands in the American West.[17]

The sharp divide in Congress, with its presidential vetoes and government shutdowns, reflected harsh rhetoric that persisted outside of Washington. Economists, systems analysts, and ecologists continued to argue over resource scarcity, population growth, and humanity's place on the planet. In the early 1990s, leading authors from the 1970s updated their earlier reports with resounding affirmations of their earlier predictions. In *The Population Explosion,* Paul and Anne Ehrlich revisited the conclusions of *The Population Bomb* and argued that the preceding two decades had proven them right. "Then the fuse was burning; now the population bomb has detonated." They asked plaintively, "Why isn't everyone as scared as we are?" Humanity was "breeding itself into a corner," faced with either a sudden and violent nuclear holocaust or a protracted environmental and social collapse in which "starvation and epidemic disease" raise global death rates across the planet. The Ehrlichs again denounced American "superconsumers" who had to learn to "tread more softly on the planet." "Halting U.S. population growth," Ehrlich wrote in a 1994 essay, was "critical" for "securing Americans and the rest of the world from problems generated or exacerbated by overpopulation." Ehrlich called for restricting immigration to the United States, while pressing to reduce US family size to 1.3 children, cut consumption, and improve conditions in poor nations.[18]

For Ehrlich and many like him, entrenched positions held, now stated all the more emphatically. In *Beyond the Limits* (1992), Dennis Meadows and his coauthors of *The Limits to Growth* affirmed their original thesis, declaring that "without significant reductions in material and energy flows, there will be in the coming decades an uncontrolled decline in per capita food output, energy use, and industrial production." In the 1993 *Global 2000 Revisited*, Gerald Barney reviewed the predictions of the original *Global 2000 Report* and repeated its claims about the future: "If present beliefs and policies continue, the world in the 21st century will be more crowded, more polluted, less stable economically and ecologically, and more vulnerable to violent disruption than the world we live in now. Serious stresses involving inter-religious relations, the economy, population, resources, environment, and security loom ahead. Overall, Earth's people will be poorer in many ways than they are today." Barney appealed to the world's faith traditions to lead people to alter their beliefs and practices.[19]

Critics also renewed familiar attacks. From the left, Paul and Anne Ehrlich were assailed for serving on the advisory boards of the Federation for American Immigration Reform, Californians for Population Stabilization, the Carrying Capacity Network, and other immigration restriction organizations. With anti-immigrant sentiment growing in California and the nation, the Ehrlichs struggled to straddle an increasingly untenable political divide, advocating lower immigration while projecting a pro-immigrant attitude. In 1993, Paul Ehrlich publicly distanced himself from a report that claimed to prove that immigrants were an economic burden on society. The Carrying Capacity Network, which promoted the study, joined that year with other immigration opponents to advocate passage of Proposition 187, a controversial 1994 California ballot amendment barring illegal immigrants from a range of public services. The struggle over immigration restrictions would roil the environmental movement during the 1990s. Anne Ehrlich, who served on the Sierra Club board, found herself drawn into a harsh conflict over whether the club should take an official position on immigration and stabilization of the US population. Although she appealed for a rational and non-discriminatory policy, or for simply steering the club away from the polarizing issue, the "defenders of immigration," Anne later recalled, "thought I was a racist." In a hotly contested vote, the members of the Sierra Club decided against adopting an explicit immigration policy. Paul Ehrlich continued to attract some public protest, attacked for lending "credibility to the anti-immigrant environmental ideology." The Ehrlichs ultimately withdrew from advisory roles with the Carrying Capacity Network and Federation for American Immigration Reform because the organizations' emphasis on Hispanic immigration seemed to them "racist."[20]

Mainstream economists also struck some familiar notes in their criticism of the new round of environmental predictions in the early 1990s. *Beyond the Limits,* Yale economist William Nordhaus wrote, was just "Lethal Model 2," a retread of the 1972 *Limits to Growth.* The new report served up the "same cast, plot, lines, and computerized scenery." Nordhaus warned that the book's prescription risked "sav[ing] the planet at the expense of its inhabitants." Economic decline could result from any of the "lethal conditions" discussed in the book. But Nordhaus saw many ways around the pessimistic scenarios. For two centuries, he wrote, "technology has been the clear victor in the race with depletion and diminishing returns." Looking ahead, Nordhaus estimated that resource constraints—energy, mineral, and land scarcity, as well as global warming—would slow economic growth by approximately one-third of a percentage point per year. Limited resources therefore presented "a small but noticeable impediment," one easily surmounted by technological advance. Lawrence Summers, then chief economist for the World Bank, echoed Nordhaus in attacking the new *Limits* book, saying that the computer model had "no validity" and only fed back the authors' own views. "We are as close to 1930 as we are to 2050," Summers argued. Technological change occurred at an extraordinary pace. "Did people in 1930 anticipate computers or antibiotics?" Summers called economic growth a moral imperative to improve the lives of people in the developing world.[21]

In May 1995, Julian Simon took his critique of Ehrlich and others to a new level of provocation. Shortly after the twenty-fifth anniversary of Earth Day, Simon ventured into Ehrlich's intellectual backyard, writing an opinion essay for the *San Francisco Chronicle* that argued that Earth Day's premises were "entirely wrong." The "doomsaying environmentalists"—here he singled out Paul Ehrlich by name—had warned of famines, dying oceans and lakes, and impending resource scarcity. Simon recalled his own embattled presentation on Earth Day 1970, which he embellished as a "physical brawl" with a fellow professor. Simon declared himself vindicated by the events of the next quarter century. "On average, people throughout the world are living longer and eating better than ever before. Fewer are dying of famine than in earlier centuries. The real prices of food and other raw materials are lower. . . . The major air and water pollutions in advanced countries have been lessening." Simon declared, "Every measure of material and environmental welfare in the United States and in the world has improved rather than deteriorated. All long-run trends point in exactly the opposite direction from the projections of the doomsayers."[22]

Simon then recounted the story of his bet with Ehrlich. A "single bet proves little," Simon acknowledged. So he offered to make a new wager that "any trend pertaining to material human welfare will improve rather than get worse." No other "prominent doomsayer" had been willing to bet him, Simon claimed.

So he called out the "chief 'official' doomsayer, Vice President Al Gore." Gore's book about environmental problems, *Earth in the Balance,* was "as ignorant and wrongheaded a collection of clichés as anything ever published on the subject." Simon admitted that his own bluster and challenges made for "unpleasant, rude talk." But proposing a bet was the "last refuge of the frustrated." He complained, "After 25 years of the doomsayers being proven entirely wrong, their credibility and influence waxes ever greater."[23]

Undeterred by losing the previous bet, Paul Ehrlich jumped at Simon's offer, determined to prove his point this time. Ehrlich and a colleague, the Stanford climatologist Stephen Schneider, responded the following week in the *San Francisco Chronicle.* They denounced Julian Simon as a leader in a campaign of deception that ignored the broad scientific consensus, including the "World Scientists' Warning to Humanity," signed by seventeen hundred scientists. Ehrlich admitted to having once made the "mistake of being goaded into making a bet" with Simon on a matter of "marginal environmental importance." Although he had expressed an eagerness to make the wager in 1980, now he said that he had been "schnookered." Prices of metals did not indicate environmental quality. "This time, we are going to ram it down his throat," Ehrlich said of his determination to prove Simon wrong and, if possible, publicly humiliate him. Ehrlich and Schneider challenged Simon to a new bet on different terms. They proposed betting a thousand dollars each on fifteen indicators over the next decade. The indicators included measures of material change: concentrations of carbon dioxide, nitrous oxide, and ozone in the atmosphere, global temperatures, tropical forest area, and quantities of fish, rice, and wheat per person.[24]

Simon refused Ehrlich and Schneider's terms. Their proposed indicators affected human welfare only indirectly, Simon argued. Simon instead suggested indicators that directly measured human health and economic well-being, things like life expectancy, leisure time, purchasing power, and commodity prices. Rather than bet on change in a physical entity such as the ozone layer, Simon suggested measuring "the trend in skin cancer death." The physical world, Simon argued, could change around us, but progress of human society would continue, bolstered by new technologies, adaptation, and markets.[25]

Ehrlich called Simon "clueless." Simon did not "understand anything about risk, trends or things that affect human welfare," Ehrlich said in an interview at the time. "We do not have the slightest little twinge of interest in debating with him." Ehrlich and Schneider rejected Simon's proposed measurements of human welfare instead of environmental change. Life expectancy depended on many factors and could also increase temporarily, before falling. "Simon wants to bet on all of what people do, good and bad, and nothing else. We deliberately focus on negative environmental or social trends because *they are the ones that need to be fixed* regardless of whether other trends are positive." Life expectancy would improve even more, Ehrlich claimed, if the negative trends that he identified turned around. Ehrlich said that he engaged in this "betting foolishness" only to get Simon to retract his claims or to get the public to see that Simon "blusters and asserts, but won't back up his own rhetoric." Ehrlich attributed corrupt motives to Simon and other optimists, saying that they clearly sought to facilitate rapid resource depletion for private gain, "leaving the rest of us and our descendants to bear the risks."[26]

Rather than find common ground, or even common terms for a second wager and debate, Ehrlich and Simon denounced each other in increasingly ad hominem and abusive language during the 1990s. "If Simon disappeared from the face of the Earth, that would be great for humanity," Ehrlich told a reporter from the *Wall Street Journal* in 1995. "I already know I'm a jerk," Simon countered. "But I've been right every time. I'll be right this time." Ehrlich, Simon said, was "shameless," "arrogant," and an "unblemished failure." Both men also criticized journalists and other public figures for manipulating facts and ignoring the logic of their side. The titles of two of their books from this period—Simon's *Hoodwinking the Nation* and Ehrlich's *Betrayal of Science and Reason*—convey their increasing frustration and bitter tone. Reviewing Al Gore's *Earth in the Balance* in 1992, for example, Simon wrote that "truth" was under siege, "rather than our very durable planet." Gore's book was just an "ignorant . . . collection of cliches." Simon considered Gore's ignorance "willful rather than naive." Gore simply chose to "ignore the scientific literature." Simon particularly condemned Gore for his embrace of an Ehrlich-like environmental agenda when Ehrlich's predictions over the previous two decades had invariably been wrong. In another essay, Simon complained, "Every statement I made in 1970 about the trends in resource scarcity and environmental cleanliness turned out to be correct." Yet the environmental agenda promulgated by national and international institutions remained unchanged. Simon thought that there was a war on the truth and a campaign against economics. His critics emphasized their "negative assessment of humanity's situation, when in fact everything is coming up roses." Much of what environmentalists call "waste" is just the "necessary byproduct of living." Simon took to attaching red plastic devil horns to his bald head for public appearances to underscore how he was vilified by environmentalists.[27]

Paul Ehrlich remained Julian Simon's chief nemesis. Simon

Julian Simon teaching a business class at the University of Maryland and wearing
the plastic red devil's horns that he often wore for public talks.
Courtesy of the family of Julian Simon.

repeatedly attacked Ehrlich for his ideas about population control and scarcity, often returning to Ehrlich's original formula-
tions in the 1968 *The Population Bomb*. Simon refused to back
down from his idea that resources were essentially infinite. In a
1997 essay, Simon acknowledged the pressures to change his
stance: "Well-wishers have advised me to 'admit' that resources
are limited to the capacities of the planet, thinking that this
will keep me from 'losing credibility.'" But Simon insisted on
standing his ground. "I must seem pigheaded," he acknowledged. He would not back down on his theoretical point.[28]

Ehrlich and his colleagues accused Simon and others of
a "stream of misinformation." Ehrlich called it "brownlash"—
defined as "the deliberate attempt to minimize the seriousness
of environmental problems through misuse or misreporting
of science." The "brownlashers," Ehrlich said, always want to

"build exemptions . . . from the laws of nature." They did not understand the basic laws of thermodynamics concerning entropy and the conservation of energy. Nor did they understand the simple mathematics of exponential growth. In a fit of hyperbole, Simon had once written, "We now have in our hands—in our libraries, really—the technology to feed, clothe, and supply energy to an ever-growing population for the next 7 billion years." This was clearly ridiculous—or, at the very least, hard to prove. In seven billion years, Ehrlich wrote with exasperation, a population growing at just one-millionth of the present growth rate would yield "more people than there are elementary particles in the universe!" Norman Myers, a biologist ally of the Ehrlichs, asked, "Would Simon's arithmetic be permitted of a freshman student?" Myers defended Ehrlich more broadly against critics such as Simon who questioned Ehrlich's predictions of mass starvation. "Where, [Ehrlich's] critics demand, have these starvation calamities occurred? During the years since Ehrlich's warning, at least 250 million and perhaps as many as 500 million people have died of hunger and hunger related diseases. . . . How many deaths do his critics want before they are satisfied that Ehrlich was correct?" Myers's appeal to the evidence of hunger-related deaths and diseases was impossible to verify, however, due to the uncertainty of famine statistics. Although many millions of people have died of hunger-related causes since the late 1960s, the connection between famine and population size is tenuous. Historians of famine emphasize instead dysfunctional markets, oppressive states, and warfare as the primary causes of famine-related deaths.[29]

Attacks by Julian Simon and other conservative critics did not diminish Paul Ehrlich's professional stature, particularly among scientists and liberal philanthropists. To the contrary,

the 1990s provided many opportunities for scientists and environmentalists to celebrate the Ehrlichs' work. Starting in the
1980s, Paul Ehrlich had received distinguished achievement
and service awards from the Sierra Club, World Wildlife Fund
International, and the American Association for the Advancement of Science. During the 1990s, as Paul and Anne entered
their sixties, Paul received many of the top awards available to
him, sometimes jointly with Anne. The Crafoord Prize from
the Royal Swedish Academy of Sciences and a MacArthur Fellowship in 1990, the Volvo Environment Prize in 1993, the UN
Environment Programme Sasakawa Environment Prize in
1994, a Heinz Award for the Environment in 1995, the Heine-
ken and Tyler Prizes in 1998, and Blue Planet Prize in 1999.
All told, the Ehrlichs received more than a million dollars in
prize money during the 1990s, a significant portion of which they
plowed into the Population Biology Gift Fund that supported
their work at Stanford. Some of the prizes, such as the Crafoord
Prize, emphasized Paul Ehrlich's scientific contributions to
conservation biology and the ecology and genetics of animal
populations. "If Paul Ehrlich is not deemed 'eminent,' who
among our community would be?" Ehrlich, the Ecological Society of America declared, had "revolutionized our thinking about
population biology, density dependence, and co-evolution."
Others, such as the Heinz Award, focused on his and Anne's
environmental leadership and service to humanity for warning
of the "dangers of ecological carelessness and arrogance." The
Ehrlichs had been subject to "harsh criticism," the Heinz
Award citation noted, but had accepted the criticism "with
grace as the price of their forthrightness." Attacks on Paul and
Anne's work thus marked their purity and toughness. For Paul
Ehrlich, this external validation was confirmation that he was
right. It reinforced the self-confidence, and even arrogance, that

allowed him to denounce his critics as "morons" and "idiots." Since the elite scientific community sided with him, he should not have to waste his time dealing with people like Julian Simon.[30]

By contrast, riffing on Ehrlich's MacArthur Fellowship, Julian Simon joked to a reporter from *Wired Magazine* in 1997, "I can't even get a McDonald's!" Simon enjoyed his newfound celebrity and the spreading influence of his ideas. But he also complained that the economics profession treated him poorly while his opponents still clung to power. Writing personal notes to himself in 1988, Simon said that he "*thought* he had hit the jackpot" in 1981, but things turned out to be more complicated. His professional work "had a large effect in changing the thinking of both academic researchers and the lay public." Yet the economics profession did not "take me to its bosom on this account" or encourage his future professional work. Simon was not invited to join any boards of economics journals, nor was he invited to give talks at prominent economic meetings. He complained that he did not get the credit that he deserved for innovations in statistical resampling methods and techniques for "bootstrap economics." Simon felt similarly unappreciated at the University of Maryland, feeling that he didn't get a salary increase in his last ten years because people didn't like his controversial ideas about population and immigration. While this may not have been literally true—his salary increased from $72,893 in 1988 to $93,428 in 1997—adjusted for inflation, his salary diminished in value, faring slightly worse than average faculty salaries during this period. The salient point is that Simon felt unappreciated, and considering his public profile and his many publications, he clearly was not valued highly. Simon struggled constantly for basic levels of university secretarial and administrative support for his research and public activities, all the more important after he parted ways with the Heritage Foundation.[31]

In the public realm, Simon thought that he had successfully eroded the scientific basis for the individuals and organizations that opposed his viewpoints on population. Yet his opponents continued to dominate public thinking. He reflected, with some despair, that "though I may have made a dent in the armor of the opposing viewpoint . . . the opposing viewpoint will continue to roll on inexorably, though perhaps with a bit less exuberance and carelessness than in the past." In a letter to a mentor at the University of Chicago, Simon complained of the "pain of expected rejections, which come with greater frequency and greater surety nowadays from professional journals." He struggled to understand the tepid or negative reaction to his work, asking "why I get the reactions that I do, and whether I have any basis to hope for anything different." Simon's professional isolation and his feeling that he had failed to change the population debate led to new bouts of "pain and frustration," bringing him "to the brink of depression many times." Simon fended off a slide into darkness each time, but with difficulty. From this vantage point, Simon's wearing of the devil's horns in public seems as much a coping mechanism for handling attacks as a self-assured provocation aimed at his critics.[32]

In 1989 and then again in 1992, Simon also experienced health incidents, either minor strokes or a severe food allergy that left his left side numb. Regardless of the cause, the incidents, as well as the death of several good friends, left him thinking about his mortality, and about what he had accomplished in his life. The prospect of death seemed to set off deepening self-doubt about the futility of his work. A new sense of urgency drove him to try to complete unfinished projects. As part of his effort to gain credit for his contributions, he drafted a book in the early 1990s, which he never published, on the volunteer auction plan for airline bumping. He also tried to correct the record about his resampling methods. "I am like a fellow trying to push eight peanuts up a hill with his nose," he wrote in 1994 about his efforts to get many unfinished projects into print. Simon also started writing an autobiography.[33]

While Simon made substantial progress on the book manuscripts, including the autobiography, he did not see several of them through to completion. On Friday, 6 February 1998, Julian Simon returned from Spain, where he had received an honorary degree from the University of Navarre. Rita and Julian decided to skip Friday night dinner at home and to go out to have a special chance to talk. Then they spent a quiet Shabbat at home on Saturday. On Sunday morning, Julian went outside on their deck to jump rope. He started to have trouble breathing. Rita told him to stop jumping rope and to relax and watch a video. Moments later the phone rang with news from Judith that she was pregnant with their first grandchild. Rita raced downstairs to tell Julian. She found him lying dead on the living room floor in front of the television. He had died of a heart attack just a few days before his sixty-sixth birthday.[34]

In obituaries in newspapers from around the country, Julian Simon was remembered as "the doomslayer," "Chicken Little's Enemy," "prophet of plenty," and a "Warrior in a 200-Year War" with Malthus. Stephen Moore, the founder of the Club for Growth and Simon's former student and acolyte, said that Simon had "routed nearly every prominent environmental scaremonger of our time: from the Club of Rome, to Paul Ehrlich, to Lester Brown, to Al Gore." John Tierney, the *New York Times* journalist and friend of Simon, commemorated Simon's passing, mocking the MacArthur Foundation for giving "genius grants" to Ehrlich and John Holdren, the losers in the bet, while ignoring Simon.[35]

After his death, Julian Simon became an icon for the free market wing of the conservative movement. Fred L. Smith, president of the Competitive Enterprise Institute, dedicated an institute publication, *Earth Report 2000: Revisiting the True State of the Planet,* with a lengthy appreciation of Simon. Smith cast Simon as a heroic soldier in the battle against the liberal establishment, who succeeded because his "brilliance forced people to listen." Julian Simon, Smith concluded, struck "some of the first blows in the effort to free man from the Malthusian miasma." The Competitive Enterprise Institute named an annual award after Simon. Stephen Moore, from the Club for Growth, received the first Julian Simon Memorial Award. In subsequent years, the award went to Robert L. Bradley, a free market energy scholar, and Bjørn Lomborg, a Danish environmental contrarian. In the fourth year, 2005, the award went to Barun Mitra, president of the Liberty Institute and a new Julian Simon Center in New Delhi, which promoted privatization and free markets. Mitra called for private conservation efforts to save the tiger; commercializing tigers could give them value and save them. "Tiger-breeding facilities," Mitra argued, "will ensure a supply of wildlife at an affordable price, and so eliminate the incentive for poachers and, consequently, the danger for those tigers left in the wild." Julian Simon had been a leading skeptic of mainstream environmental thinking. Now others carried forward his legacy. "Next step: Simonize the Global Warmists," wrote conservative commentator Ben Wattenberg in an obituary appreciating Simon.[36]

Bjørn Lomborg's controversial book *The Skeptical Environmentalist,* published in English in 2001, devoted itself explicitly to expanding on Simon's forecast for continued improvement in human welfare. In 1997, Lomborg recounted, he had read a profile of Simon in *Wired Magazine*. Simon's sharp critique of the environmental doomsday narratives intrigued Lomborg, a statistics professor in Denmark who had a long-standing interest in environmental problems. According to his narrative of self-realization, Lomborg enlisted his students to prove that Julian Simon was just a right-wing American propagandist. To their surprise, he said, they found merit in Simon's arguments. "The world is not without problems," Lomborg concluded in *The Skeptical Environmentalist*, "but on almost all accounts, things are getting better." Lomborg attacked American environmental leaders like Al Gore for suggesting that "modern industrial civilization" was "dysfunctional" and "colliding violently with our planet's ecological system." Lomborg countered, "*We have more leisure time, greater security and fewer accidents, more education, more amenities, higher incomes, fewer starving, more food and a healthier and longer life*. This is the fantastic story of mankind, and to call such a civilization 'dysfunctional' is quite simply immoral." Echoing Julian Simon's 1980 breakthrough article, "An Oversupply of False Bad News," Lomborg blamed the exaggeration of environmental threats on the news media's constant search for stories about worsening problems and on the public's confused attitudes toward risk. He did not propose that societies should just congratulate themselves. Lomborg differed from Julian Simon in his call to fight poverty and hunger in the developing world and to improve health in the developed world. Lomborg emphasized mundane priorities. In developing countries, clean water and anti-malaria campaigns. In industrialized countries, "we need to quit smoking, avoid fatty foods, get more exercise, and as a society we need to achieve a whole series of social and educational improvements."[37]

Lomborg acknowledged the threat of global warming, and he attributed climate change to human actions, such as the burning of fossil fuels. But he argued that people could adapt to a changing climate; "no ecological catastrophe" loomed around the corner "to punish us." Lomborg questioned the costs of aggressively countering climate change. Proposals to sharply and quickly curb emissions, he warned, risked extraordinary damage to the global economy—damages roughly equivalent to the economic threat of global warming itself. The "catastrophe" of global warming, Lomborg argued, lay in "spending our resources unwisely on curbing present carbon emissions at high costs." In a provocatively titled New York Times essay—"The Environmentalists Are Wrong"—Lomborg argued that for the same price of rapidly cutting carbon emissions, "we could provide every person in the world with clean water," thereby saving two million lives each year and preventing five hundred million incidents of severe disease. He urged substantial investment in new low-carbon energy technologies instead of mandatory emission cuts.[38]

Perhaps part of Lomborg's success gaining public attention lay in his ambidextrous political positioning—he leaned both to the left and right, simultaneously demanding expanded international development assistance while dismissing the need for emissions controls and environmental regulation. Whatever the cause, Lomborg's Simon-like contrarian assault on the popular "litany" of environmental disaster met with rave reviews in the Economist and other mainstream publications. This enthusiastic embrace in turn sparked a furious backlash from scientists and environmentalists. Leading scientists castigated Lomborg for what they saw as his selective use of statistics and his misreading of scientific evidence. Edward O. Wilson called The Skeptical Environmentalist a "scam" characterized by

"willful ignorance, selective quotations, disregard for . . . genuine experts, and destructive campaigning." Wilson said that
Lomborg's estimates of extinction rates were an order of mag-
nitude smaller than the most conservative estimates of author-
ities in the field. In a review in *Nature*, the biologists Stuart
Pimm and Jeff Harvey attacked Lomborg for being a new Julian
Simon. They compared Lomborg's book to a "bad term paper"
and the author's polemical style to that of a Holocaust denier.
Lomborg's critics, including Paul Ehrlich, demanded that Cambridge University Press convene an academic panel to identify
scientific errors in the book, and they asked the press to stop
publishing the book. In Denmark, a committee concluded that
Lomborg's book displayed "scientific dishonesty" (although the
committee's ruling itself was censured by another government
agency six months later). At a bookstore in Oxford, England, a
climate activist underscored the international polarization of
environmental politics by throwing a cream pie in Lomborg's
face "for lying about climate change."[39]

Some scholarly commentators noted that Lomborg and his
critics proceeded from different sets of values, and that scientific data alone could not provide the answers to fundamentally
non-scientific policy questions. In the journal *Environmental
Science and Policy,* several scholars argued that the controversy
over *The Skeptical Environmentalist* showed how science was
increasingly being used as a "trump card" in "disputes about
values." The historian of science Naomi Oreskes observed that
both Lomborg and his critics appealed to an unattainable ideal
of scientific proof that misrepresented how scientific consensus developed as a social process rather than as a product of pure
reason. Oreskes also pointed out that Lomborg and his opponents differed greatly in how they measured the state of the
world. Where Lomborg, like Julian Simon before him, calculated

John Harte and Paul Ehrlich (right) on a trail at Rocky Mountain Biological
Laboratory, July 2011. Photograph by the author.

success and failure largely in terms of human prosperity and
suffering, his critics, including Oreskes, explicitly rejected the
idea that human life is the "measure of all things." Oreskes
herself said that many of Lomborg's claims about human lon-
gevity and well-being were "surely right," but she rejected Lomborg's fixation on the quantitative aspects of human life. What
about non-economic reasons for environmental protection, she
said, including "moral, aesthetic, philosophical, emotional" jus-
tifications? Oreskes declared flatly that "humans do not have

the right to wipe other species off the planet, nor do those of us living today have the right to degrade or destroy resources that might add value to the lives of future humans." Oreskes's candor about her own values helped surface an important insight that too often eluded the protagonists in fights over the environment in the late 1990s and 2000s. "Many environmental claims are not so much about life's quantities as its qualities. They are about aesthetic and moral choices. They are about equity and ethics." Humans might improve their material lives by despoiling the planet, but did that mean that they should?[40]

In the years after Julian Simon's death, the contentious political split between liberal and conservative ways of thinking about the environment worsened. In the closely contested 2000 election, Vice President Al Gore, a passionate environmentalist, battled a Texas oilman, George W. Bush, for the presidency. After a disputed result, Bush gained the presidency. The stark contrast between Bush and the Clinton years showed in the first months of the new administration. President Bush and Vice President Richard Cheney embraced now widely held Republican positions attacking environmental science and environmental laws. Working with Tom "The Exterminator" DeLay, now Republican House Majority Leader, the Bush administration sought to overhaul national environmental regulation, continuing the effort begun under Ronald Reagan. Some Bush administration domestic policy advisers saw themselves as striving to achieve a better balance between the costs and benefits of environmental regulation. At the same time, Vice President Cheney chaired a national energy commission that primarily recommended expanded oil drilling, less public oversight, and increased nuclear power. While the energy task force endorsed energy efficiency measures, too, Cheney also dismissed energy conservation as merely "a sign of personal virtue," not a "basis for a sound, comprehensive energy policy." Cheney's position made sense from the perspective of Julian Simon. If energy scarcity did not pose a genuine problem, why press for energy conservation? Meanwhile defeated presidential candidate Al Gore reinvented himself as a globe-trotting crusader against climate change. *An Inconvenient Truth,* the 2006 documentary film about Gore's efforts to fight global warming, won an Academy Award, and Gore himself shared the 2007 Nobel Peace Prize. While Gore urged international action to stop global climate change, Bush and Cheney questioned whether humans were changing the climate and whether it even mattered. The clash between these two ways of thinking about environmental problems could not have been greater.[41]

CHAPTER SIX **Betting the Future of the Planet**

E xtreme voices have come to dominate American politics, and the partisan divide has deepened. Contributing to this division are profoundly different ways of seeing the world, such as the divergent perspectives of Paul Ehrlich and Julian Simon. Each had important insights to offer about science, economics, and society. But neither presented a vision that can stand alone. The history of Ehrlich and Simon's conflict instead reveals the limitations of their incompatible viewpoints. Their bitter clash also shows how intelligent people are drawn to vilify their opponents and to reduce the issues that they care about to stark and divisive terms. The conflict that their bet represents has ensnared the national political debate and helped to make environmental problems, especially climate change, among the most polarizing and divisive political questions.

Ehrlich and Simon both made contributions unacknowledged by the other. Paul Ehrlich's contribution—and that of environmental scientists as a whole after World War II—lay in the ability to reveal the deep connections between humans and nature and to show how the planet was changing. Through research and advocacy, Ehrlich and other scientists helped avert genuine ecological disasters, and show the risks of dangerous new technologies. If scientists had not raised the alarm about declining stratospheric ozone, nations never would have passed the 1987 Montreal Protocol, which phased out chemicals that damage Earth's protective cover against intense solar radiation. Scientific research on the potential impacts of thermonuclear war, as well as the hazards of radioactive waste and nuclear fallout, helped yield treaties to limit atmospheric testing and improve the handling of radioactive materials. More broadly, Ehrlich and other environmental scientists laid the groundwork for the new environmental regulatory regime established in the 1970s. The new environmental laws dramatically curbed air and water pollution in the United States.

The environmental scientists' impact came not just through legislation but also through a new consciousness. Ehrlich and his colleagues raised profound questions about the purpose of consumption and whether more really meant better. They also showed how important natural ecosystems are to human well-being. Instead of viewing wetlands simply as swamps to be drained, for example, scientists have shown how wetlands provide important wildlife habitat and also perform economically valuable tasks such as water management and purification. This new awareness of human reliance on the natural environment has been widely embraced by politicians, corporate leaders, and the public.

Julian Simon also had something important to contribute. He and many other economists argued that human creativity and market forces allow societies to adjust to changing circumstances and to expand efficiency and productivity. They helped fend off calls by Ehrlich and others to slow or halt economic growth in a manner that would have affected millions, or even billions, of people around the world. Economic analysis of market dynamics also demonstrated that government policies, including those regulating resource use and protecting the environment, entail economic costs. The deregulation of elements of the American economy that started in the late 1970s and accelerated in the 1980s led to greater competition as well as price declines in a variety of industries. While the loosening of regulation sometimes went much too far, as in the case of lax banking regulation and oversight, the pull-back in federal control over transportation, energy, and telecommunications encouraged innovation and economic growth in the United States. As with ecological studies conducted by scientists, the economists' research and data tested long-held prejudices and revealed the unintended consequences of policy proposals. Simon's own studies, for example, countered unfounded attacks on immigrants as a burden on the American economy. His argument for the economic benefits of immigration was one of many factors that cleared the way for the major 1986 law legalizing the status of millions of immigrants.

Sometimes rhetorical sparring partners hone each other's arguments so that they are sharper and better. The opposite happened with Paul Ehrlich and Julian Simon. Despite their respective strengths, both Ehrlich and Simon got carried away in their battle. The ready audience for their ideas encouraged them to make dramatic claims. Their unwillingness to concede anything in their often-vitriolic debate exacerbated critical weaknesses in each of their arguments.

Most fundamentally, human history over the past forty years has not conformed to Paul Ehrlich's predictions. By the most basic measure, human populations have continued to grow and no population collapse or broad-scale famine—caused by population outstripping food supply—has occurred. To the contrary. With localized exceptions, life expectancy across the globe has risen, as have per capita incomes. Food production has kept pace with population growth. Energy remains abundant. Higher food and energy prices in recent years suggest short-term shortages, and perhaps a long-term tightening of the market, but not catastrophic failure. The discrepancies in average health and welfare among nations have declined rather than increased. Countries around the world generally continue to improve their well-being rather than slip backward into greater poverty and suffering.[1]

The sustained population growth of the past forty years and growing human prosperity suggest that humanity has remained much further from its natural limits than Paul Ehrlich predicted. In a 1994 essay on "optimum human population size," he declared that the human population of 5.5 billion had "clearly exceeded the capacity of Earth to sustain it." Ehrlich and his coauthors said that an optimum population size for the planet would range from 1.5 to 2 billion people. Since that time, the planet has added another 1.5 billion people. So in what sense have humans "clearly exceeded" Earth's capacity? To be sure, many people suffer from poverty and malnutrition, and climate change threatens. But humanity has yet to run into the hard limits that Ehrlich predicted. Have we really degraded our resource base such that world populations will face catastrophic declines? We do not know for certain how many people Earth can support, and it is possible that humanity has already set the stage for its future demise. But that date still seems far off.[2]

One problem with Ehrlich's style of argument is that environmental pessimism often far exceeds reasonable predictions for how markets function and scarcity develops. Gloomy forecasts for soaring resource costs illustrate this common problem. Fears that petroleum would quickly rise into the hundreds of dollars per barrel led to another humiliating betting defeat, in this case over oil prices, for those (like Ehrlich) who believe in impending scarcity. In 2005, investment banker Matthew Simmons bet journalist John Tierney and Rita Simon (Julian's widow) five thousand dollars that oil prices would more than triple from around sixty-five dollars to a 2010 annual average of more than two hundred dollars per barrel. But the 2010 price averaged just eighty dollars. Adjusted for inflation, oil prices increased less than 10 percent over the five-year period, nowhere close to Simmons's dire forecast. Bleak, but flawed, forecasts such as these for the fossil fuel economy have broader implications because they have been the basis for disappointing national "green jobs" and "green energy" programs in the United States. Government policies have an important long-term role to play in shifting the United States away from fossil fuels toward solar and wind energy and energy efficiency. In the short term, however, dour predictions that key resources, such as solar-grade silicon, would grow increasingly scarce, and that fossil fuel prices would rise substantially, led to overly ambitious business plans and exaggerated estimates of how many new jobs would be created. Major solar power companies in the United States went bankrupt, while, at the same time, the green jobs economic programs did relatively little to stimulate job creation and spur short-term economic recovery. Aggressive subsidies by the Chinese government, of course, complicated this story by also helping competing Chinese manufacturers undercut American suppliers.[3]

Simon's victory in his bet with Ehrlich drove home an important insight relevant to these energy markets: scarcity and abundance are in dynamic relationship with each other. Abundance does not simply progress steadily to scarcity. Scarcity, by leading to increased prices, spurs innovation and investment.

Efforts to locate new resources and design cheaper methods yield new technologies. New periods of abundance occur, even overabundance or a glut. Understanding this cyclical process can be vital to crafting successful public policy. Exaggerated fears of resource scarcity can easily lead to poor economic management, including stifling price controls, panicked efforts to limit production or consumption, and national investment strategies predicated on high resource prices that turn out to be ephemeral. In other words, excessive pessimism has a cost.

Ehrlich, however, remains convinced by the essential logic of his original bet. He declared in a 2011 interview that humans were on track to "destroy their life support systems" at which point "society as we know it is going to collapse." The temptation is high to see evidence of imminent decline in current events. Yet flexible markets for energy and other natural resources, as well as human ingenuity, make the grim scenarios unlikely, and certainly not predetermined. Casual predictions that "peak oil" threatens imminent social catastrophe and massive economic disruption, or that the airline industry will "cease to exist" in a few years due to high oil prices, similarly invite skepticism of environmental claims.[4]

Julian Simon and other critics of environmentalism, however, have taken far too much comfort from extravagant and flawed predictions of scarcity and doom. Simon frequently argued that problems lead to solutions that leave humanity better off than before the problem arose. But by focusing solely and relentlessly on positive trends, Julian Simon made it more difficult to solve environmental problems. He liked to point out that the air and water were getting cleaner in the United States rather than dirtier. But Simon did not acknowledge the irony inherent in this improvement. The environment got cleaner partly because warnings by environmentalists like Paul Ehrlich prompted regulatory action. After years of struggle, environ-mentalists forced manufacturers to remove lead from gasoline and paint, improving human health. They also forced changes to motor vehicle exhaust systems that reduced air pollution–related respiratory illnesses. These environmental improvements generally cost much less than critics feared and suggest that combating climate change might also cost less than feared. Julian Simon's rosy view of the future thus undermined—and continues to discourage—efforts to address environmental problems in the present. His optimism paradoxically inhibited the kinds of problem-solving market and technological innovations that produced the improvements that he celebrated.

Simon treated markets as if they were separate from society, instead of a human creation, vulnerable to our collective blind spots and limitations. Many economists espouse Simon's view that the marketplace *can* address environmental problems adequately, *if* the markets account for the external costs of economic growth. But that is a big if. Economic research in recent decades has shown how difficult it is to eliminate information gaps and free riders and to address external costs. The idea that the market can fully account for social costs has been discredited. Market failures are unavoidable. For this reason, many economists tend to favor taxes on pollution that would force economic decision-makers to factor external, social costs into their private choices. The difficult politics of imposing these levies, however, has kept measures such as carbon taxes on fossil fuels out of legislative reach.[5]

Simon's bet with Ehrlich about mineral prices did not prove that market forces will drive these prices down in the future. Nor did that single drop in price prove that Ehrlich's concerns about threats to the environment were foolish. Simon might not have won if they had not chosen for their starting date the year 1980, the end of a decade of mostly rising metal prices. Yet Julian Simon's faith in human ingenuity and adaptability seemed to know no bounds. In his more exuberant moments, Simon claimed that humanity would figure out how to make copper from other metals, tap the resources of space to support human life, and find ways to feed and support a human population that continued to grow for thousands of years. If these views of the future were more than fairy-tale visions, Julian Simon never put forth a credible view of how they could be realized. His utopian vision often resembled an inversion of Ehrlich's dystopian future; both served as distractions from practical policies and actions.

The most pernicious current reflection of Ehrlich and Simon's clash is the ongoing political impasse over climate change. Inaccurate past claims about population growth and resource scarcity—such as Ehrlich's forecast for massive famines due to food scarcity in the 1970s and 1980s—undermined the credibility of scientists and environmentalists advocating action on climate. "By repeatedly crying wolf," the conservative judge Richard Posner wrote of Paul Ehrlich, "he has played into the hands of those who consider environmentalism a lunatic movement." Conservative commentators have warned of "apocalypse fatigue" and have frequently cited the Ehrlich-Simon bet as evidence that environmentalists are misguided fear-mongers. "Climategate did not begin with climate," the free market energy scholar Robert Bradley Jr. wrote in a 2009 essay about a controversy over climate science. Bradley tied "climate alarmism" to the "neo-Malthusianism" of the 1970s. The "doom merchants were uber-confident and still are loath to admit they were ever wrong," he argued. Conservative politicians' sustained attack on climate science draws its energy partly from decades of suspicion of environmental scientists. Conservatives who questioned Ehrlich's earlier dire claims have argued that climate warnings are just a new liberal strategy to expand government regulation and taxation. With popular support for government economic planning diminished, argued Fred Smith, president of the Competitive Enterprise Institute, the "new Malthusians" now used the pretext of an endangered, fragile earth to cover up a power grab.[6]

At the furthest extreme, these suspicions of an environmentalist conspiracy have led some prominent conservatives and Republican politicians to reject the science of climate change as a liberal hoax. "Global warming is the mother of all environmental scares," wrote the political scientist Aaron Wildavsky dismissively in a frequently quoted 1992 essay. James Inhofe, a leading congressional opponent of national climate change legislation, has urged his colleagues to "reject prophets of doom who peddle propaganda masquerading as science in the name of saving the planet from catastrophic disaster." Rush Limbaugh, the conservative radio commentator, similarly has attacked "environmental wackos" for "pure politics disguised as science." Funding from fossil fuel companies and other opponents of climate legislation has amplified these dismissive viewpoints, sowing public uncertainty about the strong scientific consensus that greenhouse gas emissions are warming the planet.[7]

Climate change, to the best of our scientific knowledge, is happening, and much of the recent global warming that we have seen appears caused by human actions. And climate change is a significant problem that threatens heavy economic and social costs. The world that humans are creating—with an increased likelihood of more intense storms, prolonged droughts, and profound changes to ecological systems—is not likely to bring changes that people will want. These are some of the vital insights of environmental scientists like Paul Ehrlich. At the same time, predictions that "billions of us will die" by the end of the century as a result of climate change or that civilization will collapse reenact the least helpful elements of Ehrlich-style environmentalism.[8]

What often gets lost in the climate debate are the lessons of the clash between Paul Ehrlich and Julian Simon. There is a serious and significant discussion to be had over what policy actions to take, and when. How much will the impacts of climate change cost, and how urgent is the need for immediate action? There are two dramatically different versions of the future. Should we count on technological innovation and economic growth to help societies meet this new challenge and adapt to change? Or must we cut emissions immediately and transform our societies in a dramatic way? The competing viewpoints echo positions held by Ehrlich and Simon. Both tend to exaggerate the consequences of their opponents' position: how expensive and disruptive it would be to shift away from fossil fuels, on the one hand, and whether it would be possible for humanity to adapt to a warmer world.

Whether human beings can increase their numbers and continue to survive on a warmer planet is only one way to assess the future. People have placed unprecedented demands on Earth's ecosystems in the past century. To adapt to population and economic growth, as well as future climate change, we are rapidly transforming the planet in the ways that Ehrlich has lamented. Even if Julian Simon proves right that humans can adapt and prosper on this changing planet, will the world that survives accord with how people want to live? The rate of resource consumption cannot be sustained without deeply altering the planet. The quest for resources also will profoundly reshape human societies. Are the risks and unequal burdens of this change acceptable? These are the questions we should be pondering, long and hard, together. Yet Ehrlich and Simon's stark framing of the future as either apocalyptic or utopian makes the conversation almost impossible.

The clashing insights of Ehrlich and Simon are necessary to help frame our thinking about the future. Our task is not to choose between these competing perspectives but rather to find ways to wrestle with their tensions and uncertainties, and to take what each offers that is of value. Ultimately, humanity's course will be determined less by iron laws of nature or by un-bounded market powers, Ehrlich and Simon's dueling lode-stars, and more by the social and political choices that we make. Neither biology nor economics can substitute for the deeper ethical question: What kind of world do we desire?

Notes

Abbreviations

AHE   Anne Howland Ehrlich
EP   Paul and Anne Ehrlich Papers, Special Collections,
        Stanford University
JLS   Julian Lincoln Simon
JSP   Julian Lincoln Simon Papers, American Heritage Center,
        University of Wyoming
PRE   Paul R. Ehrlich
RJS   Rita J. Simon

Introduction

1. Richard M. Nixon, "State of the Union Address," Washington, DC, 22 January 1970; Executive Director's Report to the Board of Directors Meeting, 13 April 1970, Saint Francis Hotel, San Francisco, EP, ser. 4, box 1, folder 1.

2. Tierney, "Betting on the Planet." JLS wrote similarly to Harold Barnett in 1977, "Every time I heard Paul Ehrlich on Johnny Carson's show, the conversation went something like this: 'Professor Ehrlich, can you explain the relationship of population and food?' Professor Ehrlich: 'It's really very simple . . . .'" JLS to Harold J. Barnett, 25 May 1977, box 9, folder 4, Harold J. Barnett Papers, 1949–1984, Collection No. 07818, American Heritage Center, University of Wyoming.

3. JLS and Gardner, "World Food Needs and 'New Proteins,'" 520–521.

4. Tierney, "Betting on the Planet."

5. PRE and AHE, *End of Affluence*, 256; PRE, "Cheap Nuclear Power Could Lead to Civilization's End."

6. Tierney, "Betting on the Planet"; JLS, "Value of Avoided Births to Underdeveloped Countries," 67.

7. Jimmy Carter, "Address to the Nation on Energy," Washington, DC, 18 April 1977, http://millercenter.org/president/speeches/detail/3398.

8. Ronald Reagan, "Announcement for President," New York, 13 November 1979, http://www.reagan.utexas.edu/archives/reference/11.13.79 .html.

9. Malthus, *Essay on the Principle of Population;* for a discussion of Malthus's views on population compared with other classical economists, see E. A. Wrigley, "The Limits to Growth: Malthus and the Classical Economists," in Teitelbaum and Winter, *Population and Resources in Western Intellectual Traditions,* 30–48; for the influence of Malthus on Darwin and Wallace, see, e.g., Barlow, *Autobiography of Charles Darwin,* 120; Wallace, *My Life,* 232, 240, 361–362; and Worster, *Nature's Economy,* 149–169.

10. Godwin, *Of Population,* 1, 449; Friedrich Engels, "Outlines of a Critique of Political Economy," in Meek, *Marx and Engels on Malthus,* 57–58.

11. Bentham, *Fragment on Government,* v; Bentham, *Introduction to the Principles of Morals and Legislation,* 1, emphasis in original; JLS, *Ultimate Resource,* 335–338.

Chapter One: Biologist to the Rescue

1. Turner, "Vindication of a Public Scholar," 51; PRE, "The Food from the Sea Myth: The Natural History of a Red Herring," presentation, Commonwealth Club of California, 21 April 1967, EP, 82-012, box 9, unmarked folder; PRE to Phyllis Piotrow, executive director of Population Crisis Committee, 2 January 1968, EP, ser. 1, box 40, folder 45; AHE to Justin Blackwelder, 13 February 1969, EP, ser. 1, box 5, folder 41; for the effort to get population into the party platforms, see William H. Draper to members of the Population Crisis Committee, 20 August 1968, EP, ser. 1, box 40, folder 45; for "joint effort" on the book, see PRE and AHE, "Population Bomb Revisited," 63; PRE to Charles Birch, 5 August 1968, EP, ser. 1, box 5, folder 42.

2. PRE, *Population Bomb,* prologue, 66–67, 169, 179; other population control advocates shared Ehrlich's view that overpopulation caused social problems, such as "crowded schools, polluted air and water, traffic jams, unspeakable slums and large families living from one generation to another in poverty." Testimony of William H. Draper Jr., Platform Committee of the 1968 Democratic National Convention Committee, Chicago, 23 August 1968, in EP, ser. 1, box 40, folder 45.

3. Wes Jackson to PRE, 13 January 1972, EP, ser. 1, box 28, folder 71; for twenty-two reprintings, see "Some Are Paying Heed . . . ," *New Scientist,* 16 September 1971, 634; PRE to Birch, 4 December 1967, EP, ser. 1, box 5, folder

42; Robert Musel, "Experts Fear World Famine in Near Future," *Los Angeles Times*, 8 December 1969; Lisa Daniel, interview by author, 10 July 2012.

4. PRE to Birch, 5 and 27 August, 6 November 1968, EP, ser. 1, box 5, folder 42; AHE to Blackwelder, 31 January 1969, EP, box 5, folder 41; AHE to Jeffrey Baker, 28 January 1969, EP, ser. 1, box 5, folder 47; Mrs. Harry A. Decker to S. Wesley Jackson and George B. Taylor, 19 May 1969, EP, ser. 1, box 28, folder 71. In 1972, Ehrlich told Jane Goodall's mother that he couldn't write an introduction to her new book because he had "nine books being either written or revised." The list of things Ehrlich had agreed to do, he said, might be "sufficient evidence to have me committed to an institution!" PRE to Mrs. Vanne Morris Goodall, 19 September 1972, EP, ser. 1, box 28, folder 56; Mary Hager, "Professor Leaps from Butterflies to Birth Control," *Washington Post*, 22 February 1970.

5. PRE, interview by author, 15 March 2010; Sally Kellock, interview by author, 10 August 2012.

6. Ruth Ehrlich and PRE, interview in *Paul Ehrlich and the Population Bomb* (Princeton, NJ: Films for the Humanities and Sciences, 1996); PRE, interviews by author, 15 March 2010, 15 July 2011; Charles Michener, interview by author, 15 November 2011; Kellock, interview; Olson, "Knowing How to Pick a Fight."

7. Kellock, interview; "List of Members of the Lepidopterists' Society, September 1948," *Journal of the Lepidopterists' Society 1948* 2 (1948); PRE, "Field Notes on the Eye Colors of the *Colias eurythemephilodice* complex; the parasitization of *Danaus plexippus;* the use of 'flyways' by *Papilio glaucus*," *Lepidopterists' News* 2, no. 8 (1948): 92; Ruth Ehrlich, interview in *Paul Ehrlich and the Population Bomb*.

8. PRE, interview, 15 March 2010; Altenberg, "From the Bedroom to the Bomb"; for the growing battle against insects, see Russell, *War and Nature*. For the broader impact of suburbanization on the environmental movement, see Rome, *Bulldozer in the Countryside*. For Maplewood in the 1940s, see Helmreich, *Enduring Community*, 65. For post–World War II suburbanization more generally, see Jackson, *Crabgrass Frontier;* and Hayden, *Building Suburbia*.

9. Olson, "Knowing How to Pick a Fight"; Osborn, *Our Plundered Planet*, vii, 32, 37, 201; Vogt, *Road to Survival*, 31, 19. The conservationist Aldo Leopold also linked population growth with global conflict in the years around World War II. See Robertson, *Malthusian Moment*, 27–28.

10. PRE, "Curriculum Vitae," Center for Conservation Biology website; PRE, interview, 15 March 2010; Olson, "Knowing How to Pick a Fight"; Michener, interview.

11. AHE, interview by author, 15 July 2011.

12. Ibid.; Daniel, interview; PRE and AHE, *How to Know the Butterflies;* Turner, "Vindication of a Public Scholar," 53.

13. PRE and Raven, "Butterflies and Plants."

14. Carson, *Silent Spring;* Linda J. Lear, *Rachel Carson: Witness for Nature* (New York: Henry Holt, 1997); Dunlap, *Faith in Nature;* Kingsland, *Evolution of American Ecology,* 179–231; Altenberg, "From the Bedroom to the Bomb." For the lack of disclosure related to radioactive isotopes like Strontium-90, see Wargo, *Green Intelligence,* 3–74.

15. EP, ser. 7, box 5, folder 9; Felix Belair Jr., "Millions May Starve in India Despite Aid, House Panel Hears," *New York Times,* 15 February 1966; Muriel Dobbin, "India Famine Plan Signed," *Baltimore Sun,* 20 April 1966; J. Anthony Lukas, "India's Famine Wiping Out Jobs on Farms," *New York Times,* 20 April 1967, 22; Arthur J. Dommen, "Potentially Rich Bihar Threatened by Famine," *Los Angeles Times,* 1 January 1967; PRE to Birch, 12 September 1966, EP, ser. 1, box 5, folder 42; PRE, *Population Bomb,* 15; AHE, interview; for the stereotypical assault on the senses, see Rotter, "Empires of the Senses."

16. PRE, *Population Bomb,* 192; PRE to Birch, 6 November 1968, EP, ser. 1, box 5, folder 42; PRE to Lester Brown, 27 January 1970, EP, ser. 1, box 40, folder 34; PRE to Ernst Mayr and Jeffrey J. W. Baker, 5 February 1969, EP, box 5, folder 47; William C. Paddock, "How Green Is the Green Revolution?" *BioScience* 20 (1970): 897–902.

17. US Census Bureau, "International Database," http://www.census .gov/ipc/www/idb/worldpop.php; "T.R.B. from Washington: Famine Is Here," *New Republic,* 18 September 1965, 6; Campaign to Check the Population Explosion, "Display Advertisement," *New York Times,* 7 January 1968; Campaign to Check the Population Explosion, "Display Advertisement," *New York Times,* 30 June 1968; Campaign to Check the Population Explosion, "Display Advertisement," *New York Times,* 12 May 1968. For a compilation of essays reflecting broad concern about overpopulation, see Osborn, *Our Crowded Planet;* for despairing warnings about the "population flood," "sea of famine," and impending starvation of hundreds of millions of people, see Snow, *State of Siege,* 22, 25, 30.

18. Robertson, *Malthusian Moment,* 91, 100; Belair, "Millions May Starve in India."

19. Hardin, *Population, Evolution, and Birth Control,* 123–125; Wade Greene, "Triage: Who Shall Be Fed? Who Shall Starve?" *New York Times Magazine,* 5 January 1975; Robertson, *Malthusian Moment,* 100; Joseph A. Califano, "Private Conscience, Public Morality," *Washington Post,* 5 December 2009.

20. Connelly, *Fatal Misconception,* 221; Cullather, *Hungry World,* 217–231; Ehrlich, *Population Bomb,* 166; PRE to James B. Plate, Little, Brown and Company, 11 December 1967, EP, ser. 1, box 5, folder 42; Charles Bartlett, "Wheat Farmers Work 1 Day in 5 for India, and That's Just Start," *Los Angeles Times,* 8 February 1966; PRE to Birch, 6 November 1968, EP, ser. 1, box 5,

folder 42; Alvin M. Weinberg, "Prudence and Technology: A Technologist's Response to Predictions of Catastrophe," *BioScience* 21 (1971): 333–334; for the counterview, see Garrett Hardin, "Not Peace, but Ecology," in Brookhaven National Laboratory, Biology Department, *Diversity and Stability in Ecological Systems*, 154 ("The problem of a starving population cannot be solved by food. We mean well in India, but we are making her situation worse").

21. PRE and Birch, "'Balance of Nature' and 'Population Control'"; PRE et al., "Weather and the 'Regulation' of Subalpine Populations," 243; PRE, "Eco-Catastrophe!" *Ramparts,* September 1969, reprinted in De Bell, *Environmental Handbook,* 161–176; PRE, "People Pollution," *Audubon,* May 1970, 4–9.

22. Frank B. Golley, "Human Population from an Ecological Perspective," in Teitelbaum and Winter, *Population and Resources in Western Intellectual Traditions,* 199–210; AHE, "Human Population—Size and Dynamics," 405; Barbara Greenberg, "Overpopulation, Finite Resources Spell Earth's Inability to Support Life," *Hartford Courant,* 26 July 1971. For the cicada story, see Bill Slocum, "A Yale Professor's Enduring Romance with Bugs," *New York Times,* 11 August 1996. See also, Eugene P. Odum, "Harmony Between Man and Nature: An Ecological View," in Meadows et al., *Beyond Growth,* 45. Legendary Yale biologist G. Evelyn Hutchinson compared humans to small crustaceans. Hutchinson warned that "we do not want an equilibrium condition analogous to that exhibited by such chronically starved Daphnia cultures in which an extreme lack of food is the fundamental agent of birth control." Hutchinson suggested that this sort of thing occurred in tropical slums, and he said that he thought the optimum human population was "considerably smaller" than the present-day numbers. G. Evelyn Hutchinson, "Some Biological Analogies," in ibid., 28.

23. George Getze, "Crisis Predicted by 1975," *Los Angeles Times,* 17 November 1967; PRE, "The Fight Against Famine Is Already Lost," *Washington Post,* 10 March 1968; Garrett Hardin, "How Freedom in a Commons Brings Tragedy," *Washington Post,* 11 May 1969; Kimmis Hendrick, "We Must Move to Spaceman Economy," *Christian Science Monitor,* 20 May 1970. Ehrlich adopted the spaceship metaphor as the organizing principle of the book he coauthored with Richard L. Harriman, *How to Be a Survivor,* and in the "rivet-popping" metaphor that frames Ehrlich and Ehrlich, *Extinction: The Causes and Consequences of the Disappearance of Species.* For a symbolic protest based on the idea of "Liferaft Earth," see "Protest!" *Sierra Club Bulletin,* December 1969, 11; for a more recent iteration, see Adam Frank, "Alone in the Void," *New York Times,* 25 July 2012.

24. Other physicists-turned-environmental scientists in the late 1960s included Stephen Schneider, who helped mediate a radical student protest at Columbia University in 1968 and went on to become a leading climate scientist; Robert Socolow, who protested the war and became an outspoken

energy expert; and Art Rosenfeld, who left physics to lead California's pathbreaking efforts in energy efficiency. Stephen Schneider, interview by author, 15 March 2010; John Harte, interview by author, 14 March 2010; see also Socolow's appeal for advice on how to "become an environmental scientist": Robert Socolow to PRE, 1 December 1969, in EP, ser. 1, box 43, loose materials.

25. John Holdren, "Introduction," in Smith, Fesharaki, and Holdren, *Earth and the Human Future*, 73, 105. Harrison Brown's influence also can be seen in PRE, AHE, and Holdren, *Ecoscience*, whose first chapter features an epigraph from Brown about the interlocking nature of population growth, resources, and agricultural production. Brown, *Challenge of Man's Future*, 105.

26. PRE and Holdren, "Population and Panaceas," 1065, 1070–1071. For an examination of the carrying capacity concept, see Sayre, "Genesis, History, and Limits of Carrying Capacity"; PRE and Holdren, "Starvation as a Policy."

27. PRE to Barry Commoner, 28 August 1970, in EP, ser. 1, box 43, loose materials; Bernard Berelson to PRE, 6 February 1970, EP, box 40, folder 45; Holdren and PRE, eds., *Global Ecology: Readings Toward a Rational Strategy for Man* (New York: Harcourt Brace Jovanovich, 1971); Paul R. Ehrlich, John P. Holdren, and Richard W. Holm, eds., *Man and the Ecosphere; Readings from Scientific American* (San Francisco: W. H. Freeman, 1971); PRE, AHE, and Holdren, *Human Ecology*, ch. 7. For an early formulation of the I = PAT equation, see PRE and John P. Holdren, "Impact of Population Growth."

28. "Director John Holdren," http://www.whitehouse.gov/adminis tration/eop/ostp/about/leadershipstaff/director, viewed 25 January 2013.

29. John Harte, interview by author, 14 March 2010.

30. Rorabaugh, *Berkeley at War*, 91–95; Shelley Mayfisher, "7:18 Train: Commuters Stand Against the War," *Yale Daily News*; "Scientists Ponder War Research; Will Hold 'Day of Reflection,'" *Yale Daily News*, 27 February 1969; "Scientists Hit ABM, Government," *Yale Daily News*, 5 March 1969; Harte, interview.

31. John Harte and Robert Socolow, "Report on Everglades," *New York Times*, 25 September 1969; "The Saving of the Everglades," *New York Times*, 18 January 1970; Robert B. Semple Jr., "Everglades Jetport Barred by a U.S.-Florida Accord," *New York Times*, 16 January 1970; Gilmour and McCauley, "Environmental Preservation and Politics"; Harte, interview; Harte and Socolow, *Patient Earth*; Harte, *Consider a Spherical Cow*.

32. Harte, interview; AHE, interview; Daniel, interview; PRE and AHE to Dear Friends, December 1989, EP, acc. ARCH-2003-076, box 1, folder 2; history of RMBL: http://rmbl.org/rockymountainbiolab/history.html; PRE, email communication with author, 11 January 2013; "School in Deserted Mining Town," *Christian Science Monitor*, 22 March 1932.

33. PRE, *Population Bomb*, xi, 135; PRE, "People Pollution," *Audubon*, May 1970, 4; Hendrick, "We Must Move to Spaceman Economy."

34. Hardin, "Tragedy of the Commons"; Hardin, "How Freedom in a Commons Brings Tragedy," *Washington Post*, 11 May 1969. PRE cited Hardin's article approvingly in PRE and Holdren, "Population and Panaceas," 1068. For environmental histories that critique Hardin's understanding of the "commons," see McEvoy, *Fisherman's Problem;* Warren, *Hunter's Game;* and Jacoby, *Crimes Against Nature.*

35. PRE and John Harte, interview by author, 15 July 2011; Executive Director's Report to the Board of Directors Meeting, 13 April 1970, Saint Francis Hotel, San Francisco, EP, ser. 4, box 1, folder 1; PRE to Bulletin of the Ecological Society of America, n.d., EP, ser. 1, box 42, folder 10.

36. PRE, *Population Bomb*, 140; PRE to Jeffrey Baker, 19 September 1968, EP, ser. 1, box 5, folder 47; PRE to Charles Wurster, 13 June 1969, EP, ser. 1, box 44, folder 59; Ehrlich, *Population Bomb*, 147–148, citing Garrett Hardin; Executive Director's Report, 13 April 1970; Wurster to PRE, 1970, 24 January 1969, EP, ser. 1, box 44, folder 59; Hoff, *State and the Stork*, 181–182; PRE, "People Pollution"; "I didn't want to go out proselytizing for a contraceptive technique without being personally able to testify to its effec tiveness, ease and lack of effect on one's sexual behavior," Abraham said. Ray Ripton, "Fear for Environment Reaches Grass Roots," *Los Angeles Times*, 15 February 1970; "Ann Landers," *Washington Post*, 25 August 1970; Clay Gouran, "A Quick Painless Answer to Overpopulation," *Chicago Tribune*, 16 August 1970; Victor Cohn, "A Male Alternative to Pill," *Los Angeles Times*, 4 October 1970; Birch to PRE, 28 January 1970, EP, ser. 1, box 5, folder 42 (including 2 February 1970 clipping from *Time*); see also Baker, "Science, Birth Control, and the Roman Catholic Church."

37. Daniel, interview; AHE, interview; AHE to Blackwelder, 31 January 1969; "Population Experts Urge Government Birth Controls to Avert Disaster," *Hartford Courant*, 3 May 1970; PRE, "The Fight Against Famine Is Already Lost," *Washington Post*, 10 March 1968; AHE to Blackwelder, 31 January 1969; Kathy Sipult, "Can Mankind Pass This Course in Survival?" *Wichita Beacon*, March 1970, EP, ser. 1, box 28, folder 71; Gordon Orians to Albert Meisel, 25 February 1970, EP, ser. 1, box 40, folder 83.

38. PRE on *Monday Conference*, ABC Television, August 1971. Getze, "Crisis Predicted by 1975," *Los Angeles Times*, 17 November 1967; PRE, *Population Bomb*, 136; Garrett Hardin, "Multiple Paths to Population Control," *Family Planning Perspectives*, 2, no. 3 (1970): 24–26; PRE to Stuart Mott, 9 January 1969, EP, ser. 1, box 40, folder 20.

39. Shirley Radl to PRE, 2 April 1970, Radl to Edgar Chasteen [1971?], EP, ser. 4, box 1, folder 1; PRE, *Population Bomb*, 138–139; PRE, "Population Control or Hobson's Choice," in Lionel Ray Taylor, ed. *The Optimum Population for Britain: Proceedings of a Symposium Held at the Royal Geographical Society, London, on 25 and 26 September, 1969* (London: Academic Press, 1970), 160–161.

40. Stanley I. Auerbach et al. to Fellow Ecologists, 30 September 1968, Lucille Harrigan to Ecologists (Auerbach et al.), 1 October 1968, EP, ser. 1, box 40, folder 83; PRE, interview by author, 15 March 2010; 1972 Republican Party affiliation confirmed via personal communication, Joyce Jahnke, deputy registrar, Santa Clara County Registrar of Voters, 6 July 2012.

41. PRE to Richard Nixon, 25 November 1968, EP, box 40, folder 106; PRE et al. to Sen. George Murphy, 7 January 1969, and Stanford University News Service, "Press Release," 8 January 1969, EP, ser. 1, box 28, folder 66; PRE to Justin Blackwelder, 20 January 1969, EP, ser. 1, box 5, folder 41; J. Brooks Flippen, *Nixon and the Environment* (Albuquerque: University of New Mexico Press, 2000), 23–24.

42. PRE to Frank Olrich, 10 October, 5 November 1969, Olrich to PRE, 28 October 1969, EP, ser. 1, box 44, folder 83; for similar attitudes toward extreme proposals that make moderate ones look more acceptable, see AHE to Blackwelder, 13 February 1969; "Gasoline Car Ban May Die in Assembly," *Modesto Bee*, 30 July 1969; and "Exhaust Controls Offered," *California Journal*, February 1970, 51.

43. Easton, *Black Tide;* "Santa Barbara Declaration of Environmental Rights," reprinted in *Congressional Record*, 91st Cong., 2nd sess., 20 January 1970, 498; Morton Mintz, "Oil Spill 'Radicalizes' Staid Coast City," *Washington Post*, 29 June 1969, 47; Char Miller and Mark Cioc, "Interview: Roderick Nash," *Environmental History*, 12, no. 2 (2007): 399–407; David Stradling and Richard Stradling, "Perceptions of the Burning River: Deindustrialization and Cleveland's Cuyahoga River," *Environmental History* 13 (July 2008): 515–535; Victor Cohn, "Lake Erie Dying, U.S. Study Finds," *Washington Post*, 3 October 1968.

44. Richard M. Nixon, "Special Message to the Congress on Problems of Population Growth," 18 July 1969, http://www.presidency.ucsb.edu/ws/index.php?pid=2132; Robertson, *Malthusian Moment*, 66–72; Hoff, "Kick That Population Commission in the Ass."

45. Richard M. Nixon, "Statement About the National Environmental Policy Act of 1969," January 1, 1970, http://www.presidency.ucsb.edu/ws/index.php?pid=2557#axzz1uP9CYtMe; Flippen, *Nixon and the Environment*, 35–36, 48–51.

46. Richard M. Nixon, "State of the Union Address (January 22, 1970)," http://millercenter.org/scripps/archive/speeches/detail/3889; Richard M. Nixon, "Special Message to the Congress on Environmental Quality," 10 February 1970, http://www.presidency.ucsb.edu/ws/index.php?pid=2757; see http://en.wikipedia.org/wiki/File:PogoEarthDayPoster1970.jpg; for the environmental as a "quality of life" movement, see Hays, *Beauty, Health and Permanence*, 22–39.

47. Other steering committee members included cochairs Pete McClosky and Gaylord Nelson and Sydney Howe, Daniel Lufkin, Charles

Creasy, Harold Jordahl, Glen Paulson, and Douglas Scott. PRE to Denis
Hayes, 24 February 1970, and Environmental Action: April 22 VI: 1, 31
January 1970, EP, ser. 1, box 44, folder 50; Charles Elwell, "High School,
College Students to Dramatize Pollution," *Los Angeles Times*, 19 April 1970;
Nan Robertson, "Earth's Day, Like Mother's, Pulls Capital Together," *New
York Times*, 23 April 1970.

   48. See, e.g., *Calvert Cliffs' Coordinating Committee v. United States
Atomic Energy Commission*, 449 F. 2d 1109 (D.C. Cir. 1971); and *Wilderness
Society v. Morton*, 479 F. 2d 842 (D.C. Cir. 1973). More generally, see Lazarus,
Making of Environmental Law; Dewey, *Don't Breathe the Air*; Milazzo, *Unlikely
Environmentalists*; and Coates, *Trans-Alaskan Pipeline Controversy*.

   49. For "Malthusian nightmare," see Richard M. Nixon, "Second
Annual Report to the Congress on United States Foreign Policy," 25 February
1971, in *Public Papers of the United States, Richard Nixon, 1971* (Washington,
DC: GPO, 1999), 334; *Environmental Action*, April 22 VI: 1, 31 January 1970;
Flippen, *Nixon and the Environment*, 69–71; PRE to Sen. George McGovern,
19 April 1972, "Draft press release," 14 October 1972, EP, ser. 7, box 5, folder
40; and James Rathlesberger, ed., *Nixon and the Environment: The Politics of
Devastation; Thirteen Essays* (New York: Taurus Communications, 1972).

   50. Flippen, *Nixon and the Environment*, 135–136, 140, 148; Richard
Nixon, "Special Message from the President to the Congress About Reorgani-
zation Plans to Establish Environmental Protection Agency and the National
Oceanic and Atmospheric Administration," 9 July 1970, http://www.epa
.gov/aboutepa/history/org/origins/reorg.html. Hoff, "Kick That Population
Commission in the Ass," 42–47.

   51. PRE to McGovern, 19 April 1972, Harold E. Seielstad to Board of
Directors, 14 December 1971, EP, ser. 4, box 1, folder 1. Ehrlich first cautioned
McGovern not to take on the abortion issue but then argued strongly that the
political landscape had changed in favor of abortion rights. As evidence, Ehrlich cited Nelson Rockefeller's veto of an effort to repeal New York's liberal
abortion law and the report of the Commission on Population Growth and
the American Future, which advocated more liberal abortion laws. PRE to
McGovern, 21 June 1972, Paul J. Growald to Robert Rickles, 6 September
1972, PRE to McGovern, 25 September 1972, Growald to John Holum, 31 May
1972, "Draft press release," 14 October 1972, EP, ser. 7, box 5, folder 40.

   52. Daniel, interview.

   53. See, e.g., Ernst Mayr to PRE and Jeff Baker, 28 January 1969, Jeff
Baker to PRE, Mayr, and AHE, 6 February 1969, EP, ser. 1, box 5, folder 47;
Eugene Odum to PRE, 29 January 1970, EP, ser. 1, box 40, folder 112; Roger
Revelle, "Review: Paul Ehrlich: New High Priest of Ecocatastrophe," *Family
Planning Perspectives* 3, no. 2 (1971): 67, 70; Birch to PRE, 13 February 1968,
PRE to Birch, 10 September, 16 October 1968, EP, ser. 1, box 5, folder 42.

   54. John Lear to PRE, 25 March 1971, Lear to PRE, 15 March, 8 April

1971, EP, ser. 1, box 23, folder 97. Lear's edits dealt with whether energy could be addressed only through population control, production efficiency, or stopping economic growth; Lear insisted that transmission efficiency and more efficient consumption also needed to be included. EP, ser. 1, box 44, folder 84; PRE to William H. Draper Jr., Population Crisis Committee, 5 November 1970, EP, box 40, folder 47; "My blood is ready for the combat!" Jeffrey Baker wrote to PRE in 1969 as he contemplated challenging Pope Paul VI on the "Humanae Vitae." Baker to PRE and Mayr, 20 January 1969, EP, ser. 1, box 5, folder 47; Wes Jackson to PRE, 13 January 1972, EP, ser. 1, box 28, folder 71; Brown to PRE, 18 August 1972, EP, ser. 1, box 40, folder 34; Holden, "Ehrlich versus Commoner," 245.

55. Commoner, *Closing Circle*, 232–235, 255; Barry Commoner, "Environment Is Not a Motherhood Issue," *New York Times*, 7 December 1971; Egan, *Barry Commoner and the Science of Survival*.

56. Holden, "Ehrlich Versus Commoner," 246–247; "Telephone conversation with Richard Lewis, PRE," 1 May 1972; Richard S. Lewis to Barry Commoner, 4 May 1972; and Sheldon Novick to PRE, 5 May 1972, EP, ser. 1, box 23, folder 77; "Nature, Man and Technology," *New York Times*, 6 February 1972; Tom Turner, "The Vindication of a Public Scholar," *Earth Island Journal*, Summer 2009, 54; Paul R. Ehrlich and John P. Holdren, "Critique: One Dimensional Ecology," *Bulletin of the Atomic Scientists*, May 1972, 16; Barry Commoner, "Response," *Bulletin of the Atomic Scientists*, May 1972, 17; The Ecologist and Friends of the Earth, "Stockholm Conference ECO," 14 June 1972, EP, acc. ARCH-2003-76, box 8, folder 1. For an overview of the Commoner-Ehrlich conflict, see Egan, *Barry Commoner and the Science of Survival*, 109–138; and Jeffrey Ellis, "On the Search for a Root Cause: Essentialist Tendencies in Environmental Discourse," in Cronon, ed., *Uncommon Ground*, 256–268.

57. Albert F. Blaustein to Board of Directors, Zero Population Growth, 8 November 1971, EP, ser. 4, box 1, folder 1; Kevles, *In the Name of Eugenics*, 168, 174, 177, 258–259. In the 2009–2010 session, the North Carolina legislature established the N.C. Justice for Sterilization Victims Foundation to compensate some of the estimated 7,600 victims of the North Carolina Eugenics Board between 1929–1977; http://www.sterilizationvictims.nc.gov/. Michelle Kessel and Jessica Hopper, "Victims Speak Out About North Carolina Sterilization Program, Which Targeted Women, Young Girls and Blacks," *Rock Center with Brian Williams*, 7 November 2011, http://rockcenter.msnbc.msn.com/_news/2011/11/07/8640744-victims-speak-out-about-north-carolina-sterilization-program-which-targeted-women-young-girls-and-blacks; Casey Bukro, "Blacks Cool Stance Toward Ecological Issues," *Chicago Tribune*, 18 June 1970; Casey Bukro, "Blacks Leave Congress on Environment," *Chicago Tribune*, 12 June 1970. Ehrlich discusses the conference in Ehrlich and Harriman, *How to Be a Survivor*, 20–21. "The Rise

of Anti-Ecology," *Time*, 3 August 1970, 44–45; Gottlieb, *Forcing the Spring*, 328.

58. PRE and Richard W. Holm to Brian M. Heald, Special Projects Editor, Holt, Rinehart and Winston, 17 May 1967, EP, ser. 1, box 23, folder 38. Ehrlich previously attacked racial categorizations in Paul Ehrlich and Richard Holm, "A Biological View of Race," in Montagu, *Concept of Race*, 153–179; Michener, interview; PRE, interview, 15 July 2011; EP, ser. 1, box 9, folder 17; PRE and Harriman, *How to Be a Survivor*, 77.

59. Bukro, "Blacks Cool Stance Toward Ecological Issues"; Bukro, "Blacks Leave Congress on Environment." "Environmental issues are really much more critical for people in minority groups," PRE said in an interview some years later, "because they're the ones, for instance, who are forced to live down wind of the power plants—the rich upwind get the power. They are the ones who will suffer first when the food shortages grow. . . . The idea that it is a luxury position is exactly wrong." Altenberg, "From the Bedroom to the Bomb"; PRE and AHE, "Population Control and Genocide," *New Democratic Coalition Newsletter*, July 1970, reprinted in Holdren and Ehrlich, *Global Ecology*, 157–159; see also PRE and Harriman, *How to Be a Survivor*, 159.

60. James Ring Adams, "Shut Up, Professor," *Wall Street Journal*, 24 June 1974.

Chapter Two: Dreams and Fears of Growth

1. "Simon to Speak at Faculty Forum," *Daily Illini*, 27 February 1970; Linda Punch, "Population Called Good," *Daily Illini*, 28 February 1970. For another skeptical view of the population crisis, see Ben Wattenberg, "The Nonsense Explosion," *New Republic*, 11 April 1970, 18–23. Wattenberg provided a counterpoint to Ehrlich's views on *The Tonight Show* on 14 August 1970.

2. JLS, *Life Against the Grain*, 259–264; "Panel Talks on Problems of Population," *Daily Illini*, 22 April 1970; Michele A. Wittig, "Problem's Cause," *Daily Illini*, 15 May 1970; Walt Harrington, "The Heretic Becomes Respectable," *Washington Post*, 18 August 1985; Judith Simon Garrett, interview by author, 2 February 2011.

3. JLS, *Life Against the Grain*, 66.

4. Edward S. Shapiro, "The Jews of New Jersey," in Cunningham, *New Jersey Ethnic Experience*, 294–313; Helmreich, *Enduring Community*, 53–54, 65.

5. JLS, *Life Against the Grain*, 68, 38, 42; RJS, interview by author, 22 July 2009; David M. Simon, interview by author, 26 January 2011.

6. JLS, *Life Against the Grain*, 51, 59, 25.

7. Aristides Demetrios, interview by author, 29 June 2011; Dan Weinberg, interview by author, 10 June 2011.

8. JLS, *Life Against the Grain*, 129–173.

9. Friedman and Kuznets, *Income from Independent Professional Practice;* Friedman and Stigler, *Roofs or Ceilings?* Friedman, *Capitalism and Freedom;* Hayek, *Road to Serfdom.*

10. JLS, *Life Against the Grain*, 42; James, *Pragmatism;* Hume, *Enquiry Concerning the Principles of Morals.* Officially, they received the Sveriges Riksbank Prize in Economic Sciences in Memory of Alfred Nobel, which was established in 1968 and awarded by the Nobel Foundation.

11. JLS, *Life Against the Grain*, 193; Rita J. Simon, interview by author, 15 June 2010; RJS, interview, 22 July 2009. RJS subsequently examined the impact of nepotism rules in RJS, Clark, and Tifft, "Of Nepotism, Marriage, and the Pursuit of an Academic Career."

12. JLS, *Life Against the Grain*, 297; David M. Simon, interview; RJS, interviews, 22 July 2009, 15 June 2010.

13. Fussler and JLS, *Patterns in the Use of Books in Large Research Libraries;* JLS, "Worth Today of United States Slaves' Imputed Wages," 113. In a 1964 essay, JLS called for a ban on cigarette advertising, contending that the health costs of smoking cost the nation far more than the economic benefits of tobacco industry. "Illinois Prof Would Outlaw Cigarette Ads," *Afro-American,* 13 June 1964, 16.

14. RJS, *As We Saw the Thirties.*

15. RJS, *Jury;* RJS, *Transracial Adoptees and Their Families;* RJS, *Immigration and American Public Policy;* RJS, *Comparative Perspective on Major Social Problems.* See, e.g., RJS, Gail Crotts, and Linda Mahan, "An Empirical Note About Married Women and Their Friends," *Social Forces,* June 1970, 520–525; RJS, Clark, and Kathleen Galway, "Woman Ph.D."; Robert Wolf and RJS, "Does Busing Improve the Racial Interactions of Children?" *Educational Researcher* 4, no. 1 (1975): 5–10; RJS, "A Survey of Faculty Attitudes Toward a Labor Dispute at Their University," *AAUP Bulletin,* June 1966, 223–224.

16. RJS, Clark, and Tifft, "Of Nepotism, Marriage, and the Pursuit of an Academic Career"; RJS, Clark, and Galway, "Woman Ph.D."; RJS, "Observations on the Function of Women Sociologists at Sociology Conventions"; RJS, interview, 15 June 2010; Katherine J. Rosich, *A History of the American Sociological Association, 1981–2004* (Washington, DC: American Sociological Association, 2005), 124; RJS, interview, 15 June 2010.

17. JLS, "Huge Marketing Research Task"; JLS, "Some 'Marketing Correct' Recommendations for Family Planning Campaigns"; JLS, "Role of Bonuses and Persuasive Propaganda in the Reduction of Birth Rates"; JLS, *Life Against the Grain,* 239.

18. JLS, "Value of Avoided Births to Underdeveloped Countries"; JLS, *Effects of Income on Fertility;* Ann Barnett, "For Love or Money," *Daily Illini,* 6 September 1975.

19. JLS, "Puzzles and Further Explorations."

20. JLS, *Life Against the Grain*, 240; see, e.g., Kuznets, "Population and Economic Growth," 171; Richard Easterlin, "Effects of Population Growth"; see also Hoff, "Are We Too Many?" 368–389; Boserup, *Conditions of Agricultural Growth;* JLS, *Life Against the Grain*, 244–245; JLS, "Positive Effect of Population Growth on Agricultural Saving in Irrigation Systems."

21. JLS, *Ultimate Resource*, 9–10. Simon tried to publish a version of his 1970 remarks in *Science*, but the publication rejected his essay. For Simon's bitter account, see JLS, *Population Matters*, 494–505. His talk was reprinted in Pohlman, *Population*, 48–62.

22. Glover and JLS, "Effect of Population Density on Infrastructure"; JLS, "Positive Effect of Population Growth on Agricultural Saving in Irrigation Systems"; JLS, "Population Growth May Be Good for LDCs in the Long Run."

23. JLS, "Welfare Effect on an Additional Child Cannot be Stated Simply and Unequivocally," 89; JLS, "Does Economic Growth Imply a Growth in Welfare?" 133; PRE, "The Food from the Sea Myth." For Ehrlich's counterperspective on the value of population growth, see also Paul R. Ehrlich and John P. Holdren, "Overpopulation and the Potential for Ecocide," in Holdren and Ehrlich, *Global Ecology*, 78 ("Where is the gain that justifies this risk?").

24. Forrester, *World Dynamics;* Meadows et al., *Limits to Growth*, 149, 199; "On Reaching a State of Global Equilibrium" (excerpts from *The Limits to Growth*), *New York Times*, 13 March 1972, 25; see also Goldsmith et al., *Blueprint for Survival.*

25. Robert Reinhold, "Warning on Growth Perils Is Examined at Symposium," *New York Times*, 3 March 1972; Editorial, "To Prosper and Live," *Christian Science Monitor*, 4 March 1972; quotation from Robert C. Townsend in Peter Passell, Marc Roberts, and Leonard Ross, "The Limits to Growth," *New York Times*, 2 April 1972; "Display Ad," *New York Times*, 28 April 1972; Anthony Lewis, "Ecology and Politics—I," *New York Times*, 4 March 1972. Lewis rejected the idea that economic growth could serve as "a path to social justice." Rather, he argued, "Growth is a cop-out, a way of avoiding the real social and moral issue of equality." While embracing the study, Lewis also cautioned that environmentalism had a potential "dark side" in its dangerous "advocacy of totalitarian remedies." Environmentalism risked "becoming a fortress for the privileged." Anthony Lewis, "Ecology and Politics—II," *New York Times*, 6 March 1972.

26. Bertram G. Murray, "Continuous Growth or No Growth," *New York Times Magazine*, 10 December 1972; see Hirsch, *Social Limits to Growth.*

27. Ralph Lapp, "We're Running Out of Gas," *New York Times*, 19 March 1972; US Congress, House, Subcommittee on Foreign Economic Policy, "Foreign Policy Implications of the Energy Crisis"; William D. Smith, "Peterson Assures Oil Industry U.S. Will Act on Energy Needs," *New York Times*, 15 November 1972; Akins, "Oil Crisis." For a sense of the conflicts that

ensued in these remote regions, see Sabin, "Voices from the Hydrocarbon Frontier"; and Sabin, "Searching for Middle Ground."

28. Coates, *Trans-Alaska Pipeline Controversy,* 249; Richard Nixon, "Statement on Signing the Amtrak Improvement Act of 1973," 3 November 1973, http://www.presidency.ucsb.edu/ws/?pid=4031; US Congress, Senate, Committee on Interior and Insular Affairs, "Highlights of Energy Legislation from the 93rd Congress, First Session," 93rd Cong., 2nd sess., 1974, 4–6; "Running Out of Everything," *Newsweek,* 19 November 1973, cover. For an overview of 1970s US oil policies, see Paul Sabin, "Crisis and Continuity in U.S. Oil Politics, 1965–1980"; Graetz, *End of Energy;* Meg Jacobs, "The Conservative Struggle and the Energy Crisis," in Schulman and Zelizer, *Rightward Bound,* 193–209.

29. Paul E. Steiger, "Collapse of World Economy Foreseen If Growth Goes On," *Los Angeles Times,* 3 March 1972; "Economists, Ecologists Hotly Debating Growth vs. No Growth," *Hartford Courant,* 11 May 1972; David C. Anderson, "A Careful Look at Growth as Suicide," *Wall Street Journal,* 17 March 1972.

30. Allen Kneese and Ronald Ridker, "Predicament of Mankind," *Washington Post,* 2 March 1972; Max Lerner, "Just Imagine! We All Can Avoid a Certain Doomsday," *Los Angeles Times,* 10 March 1972; Passell, Roberts, and Ross, "Limits to Growth"; see also Kneese and Ridker, "Predicament of Mankind." Passell and Ross expanded on their ideas about the enduring importance and possibility of growth in Passell and Ross, *Retreat from Riches.* For a critical account of the rise of systems theory and its application in the Forrester and Limits to Growth models, see Lilienfeld, *Rise of Systems Theory,* 234–246; see also Cole et al., *Models of Doom.*

31. Barnett and Morse, *Scarcity and Growth,* 6–13. For a similar view of nuclear power as "inexhaustible, sufficiently cheap, and relatively nonpolluting," see Alvin M. Weinberg, "Prudence and Technology: A Technologist's Response to Predictions of Catastrophe," *BioScience* 21, no. 7 (1971): 333–340.

32. Barnett and Morse, *Scarcity and Growth,* 6–13.

33. Nordhaus, "World Dynamics," 1169; see also Nordhaus, "Resources as a Constraint on Growth."

34. Solow, "Notes on 'Doomsday Models'"; Solow, "Economics of Resources or the Resources of Economics"; Solow, "Intergenerational Equity and Exhaustible Resources," 41.

35. Solow, "Economics of Resources or the Resources of Economics," 13; Solow, "Economist's Approach to Pollution and Its Control."

36. Solow, "Is the End of the World at Hand?" 39; Solow, "Notes on 'Doomsday Models.'"

37. Heilbroner, *Inquiry into the Human Prospect,* 22, 32, 35–40; Heilbroner, *Worldly Philosophers;* Kenneth Boulding, "The Economics of the Coming Spaceship Earth," in Jarrett, *Environmental Quality in a Growing Economy,*

3–14; Daly, "Economics as a Life Science"; Daly, *Steady State Economics;* Costanza et al., *Introduction to Ecological Economics.* For the rise of market thinking in economics and society in the 1970s, see Rodgers, *Age of Fracture* (Cambridge, MA: Belknap Press of Harvard University Press, 2011), 41–76; and Kalman, *Right Star Rising,* 227–248.

38. Economists like Charles Schultze, Lyndon Johnson's former budget director, predicted that nuclear, solar, and geothermal power would replace oil, supplying "in effect, an infinite resource base." "Growth? Or No Growth?" *Forbes,* 15 May 1974, 116–117.

39. JLS, "Does Economic Growth Imply a Growth in Welfare?" 133.

Chapter Three: Listening to Cassandra

1. US Congress, Senate, Committee on Commerce and Committee on Government Operations, "Domestic Supply Information Act: Joint Hearings before the Committee on Commerce and Committee on Government Operations," 1, 88, 91; for a similar hearing on *The Limits to Growth* before the oil embargo, see "Growth and Its Implications for the Future: Part 1," Hearing with Appendix before the Subcommittee on Fisheries and Wildlife Conservation and the Environment of the Committee on Merchant Marine and Fisheries, House of Representatives, 93rd Cong., 1st sess., May 1, 1973.

2. "Domestic Supply Information Act" hearings, 105.

3. Ibid., 273–278; in a 1988 book, *The Cassandra Conference,* PRE and Holdren embraced the Cassandra label.

4. "Domestic Supply Information Act" hearings, 279–283.

5. PRE, *Population Bomb,* 197. As one of PRE's colleagues wrote in a draft statement that criticized Pope Paul VI's "Humanae Vitae" and dismissed technology's ability to keep up with population: "If we *are* correct, then we will have acted correctly; if not, then we will have still acted correctly." "First rewrite by Baker," EP, ser. 1, box 5, folder 47; "Domestic Supply Information Act" hearings, 280; the Ehrlichs later acknowledged as "less than compelling" their use of "de-development" and "semi-development" and "population control." PRE, AHE, and Gretchen C. Daily, *The Stork and the Plow: The Equity Answer to the Human Dilemma* (New York: Putnam's, 1995), xiii.

6. Dennis Pirages and PRE, "If All Chinese Had Wheels," *New York Times,* 16 March 1972. Pirages and PRE presented a zero-sum version of global development, in which "the greed of the few will lead to disaster for all," rather than a mutually reinforcing improvement in standards of living; PRE and Harriman, *How to Be a Survivor,* 58; PRE, AHE, and Holdren, *Human Ecology,* 279; PRE and AHE, *End of Affluence,* 4.

7. PRE and AHE, *End of Affluence,* 239, 219–228.

8. Lisa Daniel, interview by author, 10 July 2012; PRE and AHE, *End of Affluence,* 240–257; see also "The Plowboy Interview: Paul Ehrlich,"

("Anybody who has any spare capital at all is a damn fool if he or she does not put away as large a store of food as he or she can afford at the present time").

9. PRE and AHE, *End of Affluence*, 256; Berry, *Continuous Harmony*, 74, 81; Schumacher, *Small Is Beautiful;* Windolf, "Sex, Drugs, and Soybeans"; Miller, "Sixties-Era Communes," 327–351.

10. Lazarus, *Making of Environmental Law; Sierra Club v. Morton*, 405 U.S. 727 (1972), 743. For a practitioner's account of the early environmental law movement, see Halpern, *Making Waves and Riding the Currents*.

11. Jimmy Carter, "For America's Third Century Why Not Our Best?" speech delivered before the National Press Club, Washington, DC, 12 December 1974, reprinted in *Vital Speeches of the Day* 41, no. 7 (15 January 1975): 215; Carter, *Why Not the Best?* 149–150.

12. J. Carter, *Why Not the Best?* 21–23, 117–120. For Carter's gubernatorial activities on natural resources policy, see Godbold, *Jimmy and Rosalynn Carter*, 196–197, 248. For a more recent survey of the Carter presidency, see Zelizer, *Jimmy Carter*.

13. J. Carter, *Why Not the Best?* 139; Godbold, *Jimmy and Rosalynn Carter*, 129, 135; Jimmy Carter, "Inaugural Address, 12 January 1971," Jimmy Carter Museum and Library, http://www.jimmycarterlibrary.gov/documents/ inaugural_address.pdf; Jimmy Carter, "The Energy Problem: Address to the Nation," 18 April 1977, http://www.politicalspeeches.net/us-politics/ jimmy-carters-address-to-the-nation-on-energy-speech.

14. Carter, *Why Not the Best?* 99–114, 127; Godbold, *Jimmy and Rosalynn Carter*, 189, 268.

15. J. Carter, *Living Faith*, 59; Jimmy Carter, "Foreword," in James Trussell Jr. and Robert Hatcher, *Women in Need* (New York: Macmillan, 1972), ix–x; J. Carter, *Remarkable Mother*, 88, 93–100; L. Carter and Spann, *Away from Home*, 83; Bourne, *Jimmy Carter*, 279–280; Jimmy Carter, "Remarks at a Reception Following a Fundraising Dinner for Senator William D. Hathaway, February 17, 1978," in *Public Papers of the Presidents: Jimmy Carter, 1978* (Washington, DC: National Archives and Records Service, 1979), 362; PRE and AHE criticized Carter for his stand on abortion in "Jimmy," *Not Man Apart* (April 1978), EP, acc. ARCH-2003-76, box 4, folder 10.

16. Godbold, *Jimmy and Rosalynn Carter*, 238–239; J. Carter, *Why Not the Best?* 117–120; J. Carter, "For America's Third Century Why Not Our Best?"; J. Carter, *Why Not the Best?* 99.

17. NOAA, National Weather Service Forecast Office, "Presidential Inaugural Weather," http://www.erh.noaa.gov/lwx/Historic_Events/ Inauguration/Inauguration.html; Glad, *Jimmy Carter*, 410–411.

18. John Kifner, "Blizzard Tightens Energy Crisis," *New York Times*, 29 January 1977, 1; Peter Behr, "This Year, Cry of Gas Shortage 'Wolf' Is Real," *Baltimore Sun*, 23 January 1977; "Carter Asks 4-Day Work Week," *Chicago Tribune*, 30 January 1977; "Nature Intrudes on Carter's Busy First

Ten Days," *New York Times*, 30 January 1977; Austin Scott and Richard Lyons, "Weather Crisis Snows Government," *Washington Post*, 29 January 1977. For Carter's failure to persuade the country to embrace his energy policy and the need for limits, see Kalman, *Right Star Rising*, 202–218.

19. Charles Mohr, "President Warns U.S. Is Probably Entering Energy Shortage Era," *New York Times*, 31 January 1977; "Carter Dresses for Cool Crisis," *Hartford Courant*, 31 January 1977; "The Text of Jimmy Carter's First Presidential Report to the American People," *New York Times*, 3 February 1977; "Nature Intrudes on Carter's Busy First Ten Days," *New York Times*, 30 January 1977; "Carter Asks Emergency Lifting of Gas Controls," *Los Angeles Times*, 26 January 1977; Jimmy Carter, "Natural Gas Legislation Remarks at a News Briefing on the Legislation," 26 January 1977, http://www.presidency.ucsb.edu/ws/?pid=7167.

20. In his second presidential debate with Gerald Ford, Carter had complained that the United States was the "only developed nation in the world that has no comprehensive energy policy." "Text of Final Debate Between Ford and Carter," *Los Angeles Times*, 23 October 1976; Karen Elliott House, "Controversy Again Follows Schlesinger as Carter Picks Him to Be Energy Chief," *Wall Street Journal*, 24 December 1976; Jon Margolis, "Environmentalists Opposed," *Chicago Tribune*, 18 December 1976; Mary McGrory, "Energy Fox Nears the Coop," *Boston Globe*, 20 December 1976; James Schlesinger Interview, Miller Center, University of Virginia Jimmy Carter Presidential Oral History Project, July 19–20, 1984, 16.

21. Schlesinger Interview, 15; "Zeroing in on Energy Waste," *Los Angeles Times*, 2 January 1977; "Carter Orders Thermostats Turned Down," *Hartford Courant*, 23 January 1977. The daytime temperature was 3 degrees lower than the 68 degrees that President Nixon had called for in 1973–1974, during the Arab oil embargo.

22. R. Carter, *First Lady from Plains*, 175; Mondale and Hage, *Good Fight*, 229; Robert Gates and Michael Gates, *From the Shadows: The Ultimate Insider's Story of Five Presidents and How They Won the Cold War* (New York: Simon and Schuster, 1996), 70.

23. Jimmy Carter, "The Energy Problem: Address to the Nation," 18 April 1977, http://www.politicalspeeches.net/us-politics/jimmy-carters-address-to-the-nation-on-energy-speech.

24. Ibid.

25. Ibid.; Jimmy Carter, "The President's Address on Energy Problems: November 8, 1977," *Vital Speeches of the Day* 44 (1977): 98–100.

26. Vietor, *Energy Policy in America Since 1945*, 259.

27. Vietor, *Contrived Competition;* Stuart Eizenstat Interview, Miller Center, University of Virginia, Jimmy Carter Presidential Oral History Project, January 29–30, 1982, 29; Schulman, *Seventies*, 125, 128; Mattson, *What the Heck Are You up to, Mr. President?*

28. PRE, AHE, and Holdren, *Ecoscience*, xiiv–xiv. AHE later recalled that *Ecoscience* was the publication to which she had made the largest contribution and that was most significant for her. AHE, interview.

29. PRE, AHE, and Holdren, *Ecoscience*, 541–620.

30. Ibid., 178, 353, 376–377, 788, 957; G. A. Norton, "Review," *Journal of Applied Ecology* 16, no. 2 (1979): 649.

31. PRE and AHE, "Population Bomb Revisited," 64.

32. PRE and Feldman, *Race Bomb*, 4–5.

33. Ibid., 41–43. PRE again addressed genetic difference and race, calling racial distinctions among humans "nonsensical," in PRE, *Human Natures*. He later acknowledged that he had been forced into expending "great effort" to refute ideas about racial superiority. PRE and AHE, "Population Bomb Revisited," 64.

34. National Center for Health Statistics, *Vital Statistics of the United States, 2003*, Vol. 1: *Natality* (Hyattsville, MD: National Center for Health Statistics, 2005), table 1.1, http://www.cdc.gov/nchs/data/statab/natfinal 2003.annvol1_01.pdf; National Center for Health Statistics, *Birth Expectations of Women in the United States, 1973–88* (Washington, DC: US GPO, 1995), table 1; Campbell J. Gibson and Emily Lennon, "Historical Census Statistics on the Foreign-Born Population of the United States: 1850–1990," Population Division Working Paper No. 29 (Washington, DC: Population Division, US Bureau of the Census, February 1999), tables 2, 14.

35. Susan J. Lowe, "Letter to the Editor: ZPG and Restricting Immigration," *Washington Post*, 24 August 1974; Lance Carden, "Zero Population Growth Expands Goal to Actual Decline in Numbers," *Christian Science Monitor*, 19 April 1973; Susan Jacoby, "Anti-Immigration Campaign Begun," *Washington Post*, 8 May 1977; "Zero Population Growth by 2008 Is Urged for US," *New York Times*, 24 June 1977.

36. PRE, Bilderback, and AHE, *Golden Door*, xv, 196, 360.

37. Ibid., viii, xv, 352, 359.

38. JLS, *Economics of Population Growth*, 486, 490, 492; see, e.g., PRE, "People Pollution," *Audubon*, May 1970, 4–9.

39. See also Goeller and Weinberg, "Age of Substitutability," 689 ("In the Age of Substitutability energy is the ultimate raw material"); JLS, *Economics of Population Growth*, 497.

40. JLS, *Life Against the Grain*, 298; JLS, *Good Mood*; JLS, *Ultimate Resource*, 9.

41. Reprint of Harvard note in "Dear Friends," September 1983, in JSP, 8282-86-07-01, box 4, Correspondence, Professional, 1983, September; David M. Simon, interview by author, 26 January 2011; RJS, interview by author, 15 June 2010; see Shula Ankary Foundation at http://www.shula.co .il/Foundation.htm.

42. JLS, "An Almost Practical Solution to Airline Overbooking"; JLS,

"Airline Overbooking: The State of the Art: A Reply"; JLS, "Wherein the Author Offers a Modest Proposal," *Wall Street Journal*, 10 January 1977; Alfred E. Kahn to JLS, 31 May 1977, JSP, 8282-86-07-01, box 4, Correspondence, Professional, 1977–1981. JLS celebrated the system's success in a 1994 essay: "The Airline Oversales Auction Plan: The Results"; JLS, *Life Against the Grain*, 294. An invigorated consumer movement brought new impetus to the idea in the mid-1970s. After Allegheny Airlines bumped the consumer advocate Ralph Nader from a flight from Washington, DC, to Hartford, Connecticut, in 1972, Nader successfully sued the airline for fraud. Nader's case, which went all the way up to the Supreme Court, increased pressure on federal regulators to address overbooking. While Nader and other critics favored a ban on overbooking, Simon argued in the *Wall Street Journal* for his proposal to allow airlines to pay volunteers to give up their seat and take a later flight. Simon surveyed waiting passengers in Chicago's O'Hare airport and found that they generally were willing to accept relatively small sums, roughly ten dollars per hour, to wait for the next flight. *Ralph Nader and the Connecticut Citizen Action Group v. Allegheny Airlines, Inc., U.S. District Court for the District of Columbia*, 445 F. Supp. 168, 10 January 1978; *Nader v. Allegheny Airlines*, U.S. Supreme Court, 426 U.S. 290, 7 June 1976; JLS, "Wherein the Author Offers a Modest Proposal"; JLS and Visvabhanathy, "Auction Solution to Airline Overbooking"; Schulman, *Seventies*, 125; Vietor, *Contrived Competition*; Kahn to JLS, 31 May 1977; JLS to Fritz Machlup, 7 July 1978, Fritz Machlup Papers, box 64, folder 22, Hoover Institution Archives; JLS, *Life Against the Grain*, 296. JLS would continue to push for full transferability of airline tickets to foster a more effective market in air transportation. JLS, "Can't Get on a Flight? There Should Be No Problem," *Wall Street Journal*, 9 April 1987.

43. Evan Maxwell, "First Hearing to Review Immigration Policy Opens," *Los Angeles Times*, 30 October 1979; Evan Maxwell, "No Short-Term Solutions to Immigration Pressures," *Los Angeles Times*, 2 November 1980.

44. Ehrlich described John Maddox's anti-environmental book *The Doomsday Syndrome*, for example, as a "curious mixture of incompetent analysis and technology-as-religion." PRE, AHE, and Holdren, *Ecoscience*, 881.

45. Ibid., 953–955.

46. Barney, "Global 2000 Report to the President and the Threshold 21 Model"; Barney, *Unfinished Agenda*, 156.

47. Barney, "Global 2000 Report to the President and the Threshold 21 Model," 127.

48. AHE, curriculum vitae, Center for Conservation Biology website, Stanford University; AHE, interview by author, 15 July 2011; US Council on Environmental Quality and Department of State, *Global 2000 Report*, 1; the internal conflict over the report persisted even after Carter lost the November

1980 election. See correspondence in Domestic Policy Staff Energy and Natural Resources (Ward: Schirmer), box 34, "Global 2000 Task Force Report," Jimmy Carter Library; "Gloomy Picture Painted of Life on Earth in Year 2000," *Los Angeles Times*, 24 July 1980.

49. Jimmy Carter, "Address to the Nation on Energy," Washington, DC, 18 April 1977, http://millercenter.org/president/speeches/detail/3398; Jimmy Carter, "Address to the Nation on Energy and National Goals," Washington, DC, 15 July 1979, online at Wooley and Peters, *American Presidency Project*, http://www.presidency.ucsb.edu/ws/index.php?pid=32596; Jimmy Carter, "Remarks at the Annual Convention of the National Association of Counties," Kansas City, MO, 16 July 1979, online at Wooley and Peters, *American Presidency Project*, http://www.presidency.ucsb.edu/ws/index.php?pid=32597.

Chapter Four: The Triumph of Optimism

1. JLS, "Resources, Population, Environment," 1432.

2. Ibid.

3. Ibid.

4. JLS, "And Now, the Good News: Life on Earth Is Improving," *Washington Post*, 13 July 1980; Steve Singer et al., "Bad News: Is It True?" 1296–1301, 1304–1305.

5. Ibid., 1305–1307.

6. Bernard Dixon, "Comment: In Praise of Prophets," *New Scientist*, 16 September 1971, 606; and PRE, "Population Control or Hobson's Choice," 161; PRE, "Environmental Disruption"; JLS, "Environmental Disruption or Environmental Improvement?"; PRE, "Economist in Wonderland"; JLS, "Paul Ehrlich Saying It Is So Doesn't Make It So," 381.

7. PRE, "Economist in Wonderland," 46.

8. Thomas D. Kelly et al., "Historical Statistics for Mineral and Material Commodities in the United States," US Geological Survey, Data Series 140, Version 2011, http://minerals.usgs.gov/ds/2005/140/.

9. Burrough, *Big Rich*, 387–405, 426–430.

10. John Harte and PRE, interview by author, 15 July 2011; "Footnotes," *Chronicle of Higher Education*, 4 May 1981, 17.

11. Ronald Reagan, "Official Announcement," New York Hilton, New York, 13 November 1979, http://www.4president.org/speeches/1980/reagan1980announcement.htm. For more on the regulation-deregulation debate during the campaign, see Alan Stone, "State and Market: Economic Regulation and the Great Productivity Debate," in Ferguson and Rogers, *Hidden Election*, 232–259. For the rhetorical shift between Carter and Reagan, see also Rodgers, *Age of Fracture*, 15–40.

12. Cannon, *Governor Reagan*, 177–178, 297–321; Tom Goff, "Cites

Challenge of Environment as Major Issues," *Los Angeles Times,* 7 January 1970; Robertson, *Malthusian Moment,* 208; Reagan, *Reagan Diaries,* 64; Reagan, *American Life,* 29, 31, 193–194.

13. Robert A. Rosenblatt, "Reagan Praises Oil Industry, Raps 'the Doomsday' Criers," *Los Angeles Times,* 17 November 1971; William Endicott, "Reagan Unveils Environmental Protection Plan: But Hits 'Doomcriers' Who Would Try to Halt Economic Development," *Los Angeles Times,* 8 April 1972; "Let's Not Starve the Coastal Panels," *Los Angeles Times,* 25 November 1973; Rasa Gustaitis, "The Fight Over 'Improving' the California Coastline," *Washington Post,* 18 August 1974; John Dreyfuss, "Criticism Mounts over Report on Environment: Cities and Counties to Get Too Much Power," *Los Angeles Times,* 23 May 1972; "Reagan Offers Program to Preserve Resources," *Los Angeles Times,* 7 April 1972; William Endicott, "Environment Measures Advance in Assembly," *Los Angeles Times,* 15 July 1971; Ronald Reagan, *Reagan, in His Own Hand,* 318–341; Robertson, *Malthusian Moment,* 208–211.

14. Edward Walsh, "NAACP Opposition to Carter Policies Cited by Reagan," *Washington Post,* 22 January 1978; Reagan, "Official Announcement"; "Report from Bill Hallmark (Portland) on their meeting with Governor Reagan in Portland, Friday, September 26, 1980," Domestic Policy Staff Energy and Natural Resources (Ward: Schirmer), box 34, "Global 2000 [1]," Carter Library.

15. Commission on Presidential Debates, "The Carter-Reagan Presidential Debate," 28 October 1980, http://www.debates.org/index.php?page =october-28–1980-debate-transcript.

16. Harte, interview; Ernest Furgurson, "Kennedy, Anderson, Carter 1-2-3 in Grading on Environmental Issues," *Baltimore Sun,* 10 May 1980; Environmental leaders to Jimmy Carter, 25 May 1979, in Name File "Speth, J. Gustave," Carter Library; Carter's note, signed "J.C.," is written on Speth to Hamilton Jordan, 16 August 1979, in Office of the White House Staff Director, Al McDonald's Subject Files, box 10, "Council on Environmental Quality—Gus Speth," Carter Library; Eizenstat to McDonald, 15 November 1979, in Office of the White House Staff Director, Al McDonald's Subject Files, box 10, "Council on Environmental Quality—Gus Speth," Carter Library; Luther J. Carter, "Gus Speth, Planning the 'Conserver Society,'" *Science* 208 (1980): 1009–1012.

17. Philip Shabecoff, "Major Environment Leaders Back Carter Re-Election Bid," *New York Times,* 28 September 1980; Ernest Furgurson, "Kennedy, Anderson, Carter 1-2-3 in Grading on Environmental Issues," *Baltimore Sun,* 10 May 1980.

18. John F. Stacks, *Watershed: The Campaign for the Presidency, 1980* (New York: Times Books, 1981), 184; Ronald Reagan, "Acceptance of the Republican Nomination for President," 17 July 1980, http://www.americanrhetoric .com/speeches/ronaldreagan1980rnc.htm; Bill Prochnau, "Top Carter Aides

Begin Attacks on Reagan," *Washington Post*, 10 October 1980; "The Environment and the Stump," *New York Times*, 22 October 1980; "Mr. Reagan v.
Nature," *Washington Post*, 10 October 1980; Cannon, *Governor Reagan*,
495–498; Drew, *Portrait of an Election*, 114, 324; Jack Nelson, "Pollution
Curbed, Reagan Says; Attacks Air Cleanup," *Los Angeles Times*, 9 October
1980; Steve Grant, "Costle Hits Reagan on EPA Issues," *Hartford Courant*,
10 October 1980.

   19. Carter, "President Jimmy Carter's Farewell Address," 14 January
1981, http://www.jimmycarterlibrary.gov/documents/speeches/farewell
.phtml; see also Brinkley, *Unfinished Presidency*, 32. Behind the scenes a small
struggle surrounded the task force charged with developing recommendations based on *The Global 2000 Report*. Gus Speth, chairman of the Council
on Environmental Quality, had pushed harder than some of his White House
colleagues wanted. They pressed him to drop specific budget numbers and
times frames, to temper criticism of the United States, and to release the task
force document under CEQ's name only. (Handwritten notes for call to Gus
Speth, in Domestic Policy Staff Energy and Natural Resources [Ward: Schir-
mer], box 34, "Global 2000 Task Force Report.") Speth wanted Carter to act
on *The Global 2000 Report* before leaving office, but Domestic Policy staffers
in the White House blocked Speth's efforts to get his ideas for action in front
of the president. Erica Ward and Tom Lambrix to Eizenstat, "Global 2000—
Attached Memo from Gus to the President," 17 December 1980, Domestic
Policy Staff Energy and Natural Resources (Ward: Schirmer), box 34, "Global
2000 [1]"; Speth to the President, "Global Resources and Environment Task
Force—Completion of Work," 15 December 1980, Domestic Policy Staff
Energy and Natural Resources (Ward: Schirmer), box 34, "Global 2000 [1]."
With a major grant from the MacArthur Foundation, Gus Speth went on to
found the World Resources Institute to study the issues raised by *The Global
2000 Report;* Jimmy Carter, "Global Resources, Environment, and Population," *American Biology Teacher* 46, no. 6 (1984): 305–309, based on an
address to Global Tomorrow Coalition Conference on 2 June 1983.

   20. Peter Behr and Merrill Brown, "1-Year Moratorium Recommended
on New Regulations," *Washington Post*, 9 November 1980, G1; David A.
Stockman and Jack Kemp, "Avoiding a GOP Economic Dunkirk," *Wall Street
Journal*, 12 December 1980, 28; William Greider, "The Education of David
Stockman," *Atlantic Monthly*, December 1981.

   21. Robert W. Merry and Dennis Farney, "New Budget Chief Pledges 2%
Cut, Wants to Restrict Loan Guarantees," *Wall Street Journal*, 12 December
1980; Graetz, *End of Energy*, 153. For an interpretation of Reagan-era
regulatory review as part of a bipartisan response to the growth in administrative regulation, see DeMuth, "Oira at Thirty," *Administrative Law Review*
63 (Special Edition 2011): 15–25. See also Sunstein, *Risk and Reason*.

   22. Short, *Ronald Reagan and the Public Lands*, 51–53; Rowell, *Green*

Backlash, 52. For Watt's religious convictions and his embrace of the Pentacostal church, see James Watt to Dear Family, n.d., Watt Family to Dear Friends, December 1965, and Watt to Milward L. Simpson, 29 July 1966, James G. Watt Papers, acc. no. 7667, AHC, box 1, folder 14. For Watt's earlier opposition to federal pollution control efforts, see Watt to A. Fred Miller, 23 May 1968, James G. Watt Papers, acc. no. 7667, AHC, box 1, folder 16.

23. Bill Prochnau, "The Watt Controversy," *Washington Post*, 30 June 1981; Reagan, *Reagan Diaries*, 64, 93; Short, *Ronald Reagan and the Public Lands*, 60; Patricia Sullivan, "Anne Gorsuch Burford, 62, Dies; Reagan EPA Director," *Washington Post*, 22 July 2004; Eleanor Randolph, "Gorsuch: 'Ice Queen' Making Her Share of Enemies Running EPA," *Hartford Courant*, 6 November 1981; Lawrence Mosher, "Move Over, Jim Watt, Anne Gorsuch Is the Latest Target of Environmentalists," *National Journal*, 24 October 1981, 13; Russell E. Train, "The Destruction of the EPA," *Washington Post*, 2 February 1982. Gorsuch supported the first round of budget cuts at EPA but later appealed to President Reagan to avert much tougher reductions sought by David Stockman at the Office of Management and Budget. Burford, with Greenya, *Are You Tough Enough?* 75–79.

24. Gottlieb, *Forcing the Spring*, 167–175; McCloskey, *In the Thick of It*, 214; Robert A. Shanley, *Presidential Influence and Environmental Policy*, 110–111. See also Jeffrey Stine, "Natural Resources and Environmental Policy," in Brownlee and Graham, *Reagan Presidency*, 233–256.

25. Reagan, *Reagan Diaries*, 201, 204, 273, 277; Duncan Haimerl, "People in the News," *Hartford Courant*, 28 May 1983; McCloskey, *In the Thick of It*, 219. For a critical contemporary account of Reagan's first-term environmental policies, see Lash, Gillman, and Sheridan, *Season of Spoils*.

26. Altenberg, "From the Bedroom to the Bomb"; PRE to G. Evelyn Hutchinson, 29 July 1983, in G. Evelyn Hutchinson Papers, Manuscripts and Archives, Yale University Library (hereafter GEHP), ser. 1, box 15, folder 248.

27. PRE to Hutchinson, 17 June 1983, in GEHP, ser. 1, box 15, folder 247; PRE to Hutchinson, 29 July 1983, in GEHP, ser. 1, box 15, folder 248; PRE et al., "Long-Term Biological Consequences of Nuclear War," 1293; PRE, Sagan, Kennedy, and Roberts, *Cold and the Dark*. After a tour of Britain focused on the threat of nuclear winter, AHE served as a commissioner for a study of the potential impacts of nuclear war on London. Subsequently, she worked with the Sierra Club to help draw attention to the problem of hazardous waste pollution on federal property, an effort that led to the passage of the 1992 Federal Facility Compliance Act. See Clarke, *London Under Attack;* and AHE, interview.

28. JLS, *Life Against the Grain*, 312–313.

29. JLS, "Scarcity of Raw Materials"; JLS, "World Food Supplies"; JLS, "World Population Growth"; JLS, *Ultimate Resource*, vii, 345–348.

30. JLS, *Ultimate Resource*, 3–20, 283.

31. Wade Green, "The Militant Malthusians," *Saturday Review,* 11 March 1972, 49; for population estimate, see US Census Bureau, "Historical National Population Estimates," 28 June 2000, http://www.census.gov/ popest/archives/1990s/popclockest.txt; JLS, *Ultimate Resource,* 10, 335–336.

32. William G. Tucker, "Mining the Mind of Man," *Washington Post Book World,* 20 September 1981, 5; JLS and Buckley, "Answer to Malthus?" 205.

33. JLS, "Economic Objections Fall Apart When Used Against Immi-grants," *Los Angeles Times,* 1 July 1980; JLS, "Immigrants—Especially Illegals—Raise U.S. Incomes," *Newsday,* 10 July 1980; JLS, "Adding Up the Costs of Our New Immigrants," *Wall Street Journal,* 26 February 1981; JLS, "How Vietnamese Immigrants Fare in the United States," *Asian Wall Street Journal,* 25 June 1981; Robert E. Taylor, "The Latino Tide," *Wall Street Journal,* 12 June 1984; JLS, "Losing Land?" *New York Times,* 7 October 1980; JLS, "Assuring the Future of U.S. Farm Land," (with Dale McLaren), *Wall Street Journal,* 10 November 1980; JLS, "The Phantom 30,000 Acres of Lost Farmland," *Champaign-Urbana News-Gazette,* 17 May 1981; JLS, "Cropland Acres and Quality Up," *Champaign-Urbana News-Gazette,* 9 June 1981; JLS, "The Farmer and the Mall: Are American Farmlands Disappearing?" *Ameri-can Spectator,* August 1982, 18–20, 40–41; Philip M. Boffey, "Will the Next 20 Years Bring Prosperity or Decline?" *New York Times,* 12 January 1982.

34. For the history of the institutions created to support conservative ideology, see Phillips-Fein, *Invisible Hands;* Sean Wilentz, *Age of Reagan,* 89–93; Teles, *Rise of the Conservative Legal Movement;* Alice O'Connor, "Financing the Counterrevolution," in Schulman and Zelizer, *Rightward Bound,* 148–168.

35. Kahn, *Thermonuclear War;* Kahn and Wiener, "Next Thirty-Three Years," 731; William D. Tammeus, "'Limits to Growth': Where Do We Go from Here?" *Kansas City Star,* 26 October 1975; Herman Kahn and T. Mitchell Ford, "Don't Expect Doomsday," *New York Times,* 3 October 1980; Kahn and Schneider, "Globaloney 2000"; population estimates from US Census Bureau, "Total Midyear Population for the World: 1950–2050," http://www .census.gov/population/international/data/idb/worldpoptotal.php. Kahn did not coin the term *globaloney,* which was used as early as the 1940s. See, e.g., Davis, "Population and Resources," 346–349; Kahn, *Coming Boom.*

36. JLS and Kahn, eds., *Resourceful Earth,* 32–37; Francis X. Clines and Lynn Rossellini, "Briefing," *New York Times,* 26 April 1982; Philip J. Hilts, "Carter-Era Report on Global Ecology Challenged in New Study," *Washington Post,* 30 May 1983; Walter Goodman, "Academic Team, Disputing Carter Study, Sees Affluent Future," *New York Times,* 20 June 1984; Ed Fuelner to Ed Meese, undated draft letter, JSP, 8282-86-07-01, box 4: Correspondence, Professional, 1980–1983; JLS to Robert Ayres, draft letter, 19 October 1982, JSP, 8282-86-07-01, box 3: Manuscript 1982, "A Second Look at 2000," explaining that Heritage had agreed to pay expenses for meetings and

honoraria; Holden, "Simon and Kahn Versus Global 2000," 343. See Philip
Lawler to Julian Simon, undated, asking S. Fred Singer to add a section to
his essay "explaining where the Global 2000 Report went wrong," in JSP,
8282-86-07-01, box 3, Manuscript 1983, Simon, J. Global 2,000 Reconsidered; JLS to David Asman, 27 June 1983, draft letter, JSP, 8282-86-07-01,
box 3: Professional File, 1983 Global 2000 Revised, indicating that each
author had been asked to close their piece by indicating where it differed
from *Global 2000.*

37. Simon and Kahn, *Resourceful Earth*, 1–2; Philip J. Hilts, "Carter-Era
Report on Global Ecology Challenged in New Study," *Washington Post*, 30 May
1983; William Tucker, "Bullish on the Future," *Washington Post Book World*,
9 September 1984, 4; Ronald Kotulak, "Bright New World Seen for Year
2000," *Chicago Tribune*, 29 May 1983.

38. Holden, "Simon and Kahn Versus Global 2000"; JLS to Joan
Wrather and Carol Rogers, 20 June 1983, in JSP, 8282-86-07-01, box 4:
Correspondence, Professional, 1983, June, thanking them for suggesting
and arranging for a press conference. Other panelists included William
Nordhaus, Tom Schelling, and Herman Daly. Robert Pindyck to JLS, 6 April
1983, in Julian L. Simon Collection, American Heritage Center, Accession
#JSP, 8282-86-07-01, box 4, folder: 1983; Bayard Webster, "Scientists Mix
Optimists and Pessimists to View Globe's Fate," *New York Times*, 26 May
1984; William E. Geist, "About New York: When 5,000 Scientists Meet for a
Weekend," *New York Times*, 26 May 1984; JLS to David Wills, 29 June 1983,
JSP, 8282-86-07-01, box 4: Correspondence, Professional, 1983, June,
proposing a "full-scale data shootout"; Tim Schreiner, "Future Is A.) Dim
or B.) Bright (Pick One)," *USA Today*, 2 June 1983.

39. Joshua Gilder to JLS, 22 August 1983; "Excerpts from Remarks by
Vice President George Bush at the Southwest Voter Registration and Educa-
tion Project Luncheon," 9 August 1983; "Address by Vice President George
Bush at the Ohio State Spring Quarter Commencement Convocation,
Columbus, Ohio, 10 June 1983"; James A. M. Muncy to JLS, 30 September
1983, JSP, 8282-86-07-01, box 4: Correspondence, Professional, 1983,
September. For an earlier discussion of space as an "infinite" source of
resources, see Jack D. Salmon, "Politics of Scarcity Versus Technological
Optimism," 701–720. James Watt also liked what JLS had to say and invited
him to brief his staff on global environmental issues. Watt to JLS, 24 August
1983, JSP, 8282-86-07-01, box 4: Correspondence, Professional, 1983,
August; William Butz to JLS, 13 September 1983, JSP, 8282-86-07-01,
box 4: Correspondence, Professional, 1983, September.

40. Draft letter to friends, undated, JSP, 8282-86-07-01, box 4:
Correspondence, Professional, 1980–1983; JLS to Murray Polakoff and
Rudolphe Lamone, 30 June 1983, JSP, 8282-86-07-01, box 4: Correspon-
dence, Professional, 1983, June, noting the "cliffhanging aspect of my

appointment"; JLS to Aaron Wildavsky, 6 May 1983, in JSP, 8282-86-07-01, box 3: Professional File, 1983 Global 2000 Revised; Judith had skipped a year, so she and David went at the same time. Judith Simon Garrett, interview by author, 2 February 2011; JLS to friends, 1983, JSP, 8282-86-07-01, box 4: Correspondence, Professional, 1983, September.

41. David M. Simon, interview by author, 26 January 2011; JLS to John Burness, 30 September 1983, JSP, 8282-86-07-01, box 4: Correspondence, Professional, 1983, September.

42. JLS to Scott Armstrong et al., 30 September 1983, JSP, 8282-86-07-01, box 4: Correspondence, Professional, 1983, September; JLS to Burt Pines, 25 October 1983, JSP, 8282-86-07-01, box 4: Correspondence, Professional, 1983, October; Fred L. Smith to JLS, 24 October 1983, JSP, 8282-86-07-01, box 4: Correspondence, Professional, 1983, October. JLS insisted that the census made sense for the government to do and that EPA data collection also had value, since "it is only because EPA data exists that we are now able to put the boots to false claims that the environment is getting dirtier. This is an example, I think, of where the truth is served by there being more rather than less government data." JLS to Fred Smith, 10 November 1983, JSP, 8282-86-07-01, box 4: Correspondence, Professional, 1983, November. For an extended explanation of the ideas of free market environmentalism, see Anderson and Leal, *Free Market Environmentalism*.

43. JLS to Mrs. Randy Engel, JSP, 8282-86-07-01, box 4: Correspondence, Professional, 1983, October; Burt Pines to JLS, 28 October 1983, JSP, 8282-86-07-01, box 4, folder: 1983; see Simon's notes on "How a Delegate Is Chosen for the World Population Conference in Mexico City, August 1984," in JSP, 8282-86-07-01, box 4: Correspondence, Professional, 1980–1983; JLS to Kenneth Dam, 20 October 1983, JSP, 8282-86-07-01, box 4: Correspondence, Professional, 1983, October; JLS, "Testimony" before US House of Representatives, Subcommittee on Census and Population, 28 June 1984, JSP, 8282-94-09-01, box 4: unfiled papers; JLS, "Myths of Overpopulation," *Wall Street Journal*, 3 August 1984; Laura Castaneda and Joanne Omang, "A Champion of Population Growth," *Washington Post*, 13 August 1984; "The Second 1984 Presidential Debate: October 28, 1984," transcript, http://www.pbs.org/newshour/debatingourdestiny/84debates/2prez3.html; JLS, "Population Growth Can Be Good for Us," *Washington Post*, 10 November 1984; James L. Buckley, *Freedom at Risk: Reflections on Politics, Liberty, and the State* (New York: Encounter Books, 2010), 173–174.

44. JLS, "Why Do We Think Babies Create Poverty?" *Washington Post*, 13 October 1985; JLS, "There Is No Crisis; Air, Water Are Safe," *USA Today*, 12 July 1984; JLS, "Does Doom Loom?"

45. Ronald Reagan, "Interviews with Representatives of Orlando, Florida, Television Stations," 2 July 1984, http://www.presidency.ucsb.edu/

ws/index.php?pid=40123#axzz1PpYSpIeZ; Robert E. Taylor, "The Latino Tide," *Wall Street Journal*, 12 June 1984.

46. JLS, "Don't Fear Job Loss," *New York Times*, 2 August 1984; JLS, *Economic Consequences of Immigration;* JLS, *How Do Immigrants Affect Us Economically?*

47. "Fake IDs Are Aliens' Passport to Jobs," *Chicago Tribune*, 27 April 1983; "The Underclass," *Wall Street Journal*, 30 May 1985; Lee Dembart, "Immigrants—Drain on US Resources," *Los Angeles Times*, 11 May 1981; "Immigration of the Fittest," *New York Times*, 31 January 1986.

48. US Senate, *Congressional Record*, 98th Cong., 2nd sess., 8 February 1984, 2275; Rep. Bill Richardson, Democrat from New Mexico, also embraced JLS's analysis, entering it into the *Congressional Record* in the House in June 1984, US House of Representatives, *Congressional Record*, 98th Cong., 2nd sess., 20 June 1984, 17251; JLS to Edward Kennedy, 28 February 1984, JSP, 8282-86-07-01, box 4: Correspondence, Professional, 1984, February.

49. Donald Collins to Richard Larry, 27 February 1984, JSP, 8282-86-07-01, box 4: Correspondence, Professional, 1984, February; the Council for a Competitive Economy opposed a crackdown on illegal immigration: JLS to Cynthia Jo Ingham, 19 May 1983, and Cynthia Jo Ingham to JLS, 5 May 1983, JSP, 8282-86-07-01, box 4: Correspondence, Professional, 1983, April, May. Heritage hosted a debate between Simon and the executive director of the Federation for American Immigration Reform, the immigration-restriction organization that Paul Ehrlich had endorsed. Draft transcript of Conner-Simon Debate, JSP, 8282-94-09-01, box 1: unfiled papers; JLS to Roger Conner, draft, undated, JSP, 8282-86-07-01, box 4: Correspondence, Professional, 1980–1983, which indicates that Heritage's Ed Feulner and Burt Pines asked JLS to respond to Conner's critique, and JLS offered to debate Conner in public.

50. John Dillin, "Leaked Report Adds Fuel to Immigration Debate," *Christian Science Monitor*, 24 January 1986; JLS, "Bring on the Wretched Refuse," *Wall Street Journal*, 26 January 1990; JLS to James L. Kilpatrick, 27 February 1984, JSP, 8282-86-07-01, box 4: Correspondence, Professional, 1984, February; JLS to Roland A. Alum, Jr., 27 February 1984, JSP, 8282-86-07-01, box 4: Correspondence, Professional, 1984, February.

51. JLS, *Economic Consequences of Immigration*, introduction n. 2 and associated text.

52. Laura Castaneda and Joanne Omang, "A Champion of Population Growth," *Washington Post*, 13 August 1984; Walt Harrington, "The Heretic Becomes Respectable," *Washington Post*, 18 August 1985; Edith Efron, *Apocalyptics;* see also Wattenberg, *The Good News Is the Bad News Is Wrong*. For rejection from *The Phil Donahue Show* for being "too abstract and intellectual" a topic, see Tim Manners to JLS, 18 July 1983, Sandra Furton to Manners, 21 July 1983, and Manners to JLS, 27 July 1983, JSP, 8282-86-07-01, box 4:

Correspondence, Professional, 1983, July; for Heritage's support, see, e.g., Simon's request for help placing an op-ed on species extinction in 1983 and his thanks for the "Detroit coup" at the AAAS meeting. JLS to Herb Berkowitz, 20 June 1983, JSP, 8282-86-07-01, box 4: Correspondence, Professional, 1983, June; immigration paper had "scored a major success," Stuart Butler to Burt Pines and JLS, 1 February 1984, JSP, 8282-86-07-01, box 4: Correspondence, Professional, 1984, February; "repeatable magic," JLS to Berkowitz, 29 February 1984, JSP, 8282-86-07-01, box 4: Correspondence, Professional, 1984, February. For the broader strategy of conservative support for public intellectuals like JLS, see Phillips-Fein, *Invisible Hands;* David A. Hollinger, "Money and Academic Freedom a Half-Century After McCarthyism: Universities amid the Force Fields of Capital," in P. J. Hollingsworth, *Unfettered Expression,* 161–184; JLS to Jack Kemp, 6 April 1988, draft, JSP, 8282-93-06-08, box 5, unfiled papers.

53. Robert Repetto, "Why Does Julian Simon Not Believe His Own Research?" *Washington Post,* 2 November 1985; Margaret Webster, "On 'The Heretic Becomes Respectable,'" *Washington Post,* 25 August 1985; JLS, *How to Start and Operate a Mail-Order Business;* Gilbert F. White to JLS, 4 November 1983, JSP, 8282-86-07-01, box 4, Correspondence, Professional, 1983, November.

54. JLS to Robert Sassone, 27 October 1983, JSP, 8282-86-07-01, box 4, Correspondence, Professional, 1983, October; JLS to Albert Rees, 8 January 1985, draft, JSP, 8282-94-09-01, box 4, unfiled papers; JLS to Samuel Preston, 1 April 1986, draft, JSP, 8282-94-09-01, box 3, unfiled papers; JLSto friends, draft, 1987/1988, JSP, 8282-93-06-08, box 5, unfiled papers.

55. National Academy of Sciences, *Rapid Population Growth;* National Research Council, *Population Growth and Economic Development,* viii; Harrington, "The Heretic Becomes Respectable."

56. The 1986 report cited approvingly Simon's argument that population growth can stimulate agricultural production and lead to improvement in transportation facilities and access to markets, as well as his 1977 conclusion that population growth did not correlate with per capita income. National Research Council, *Population Growth and Economic Development,* 4, 14–17, 25.

57. Andrea Tyree, "Review," *Journal of the American Statistical Association* 82 (December 1987): 1180.

58. Allen C. Kelley, JLS, Joseph E. Potter, and Herman E. Daly, "Review Symposium," *Population and Development Review* 12, no. 3 (1986): 569, 575; JLS to Samuel Preston, 11 March 1986, draft letter, and JLS to Preston, 1 April 1986, draft letter, JSP, 8282-94-09-01, box 3: unfiled papers.

59. PRE to Frank Press, 20 March 1986, GEHP, ser. 1, box 15, folder 247.

60. Hutchinson to PRE, 26 March 1986, GEHP, ser. 1, box 15, folder 247; Ansley Coale, "Review: Population Growth and Economic Development:

Policy Questions," *Journal of Political Economy* 95, no. 4 (1987): 887–892; Kelley, JLS, Potter, and Daly, "Review Symposium," 582, 585.

61. PRE, "Why the Club of Earth?" *TREE* 2 (May 1987): 133–135; PRE, "AIBS News," *BioScience* 37, no. 10 (1987): 757–763.

62. PRE, "Why the Club of Earth?"; PRE, interview by author, July 15, 2011; Douglas, "Paul Ehrlich: An Interview"; Stanford News Press Release: "Population Doomed to Reach 8 Billion, Club of Earth Warns," 3 September 1988, PRE to Members of Club of Earth, 9 September 1988, "Statement of the Club of Earth on Population," 3 March 1988, GEHP, ser. 1, box 15, folder 248.

63. Lance Frazer, "Twenty Years After 'The Population Bomb,'" *This World*, 28 August 1988, 14; PRE, "Why the Club of Earth?"; PRE, "AIBS News," 761.

64. "Pro People: Proposal for an Organization," 25 February 1985, JSP, 8282-94-09-01, box 1, unfiled papers. See, e.g., JLS to Burt Pines, 31 October 1983, JSP, 8282-86-07-01, box 4, Correspondence, Professional, 1983, October.

65. JLS, "Unreported Revolution in Population Economics."

66. "Talk on Population and Econ Develo for Gabriel Roth's Group, 9 October 1988 for Oct 11," JSP, 8282-93-06-08, box 5, unfiled papers; JLS, "The Population Establishment, Corruption, and Reform," in Roberts, *Population Policy*, 39–58.

67. For an overview of the modest changes that occurred in environmental law and the idea that extreme rhetoric had blocked more substantive conservative reforms, see Lazarus, *Making of Environmental Law*, 98–124.

Chapter Five: Polarizing Politics

1. Tierney, "Betting on the Planet," 81; John Tierney, "The Population Crisis Revisited," *Wall Street Journal*, 20 January 1986; John Tierney, "The $10,000 Question," *New York Times*, 23 August 2005; Julian Sanchez, "Fifth Columnist: New York Times Columnist John Tierney Brings Libertarian Ideas to America's Big-Government Bible," *Reason Magazine*, 14 September 2005; John Tierney, "Optimism by the Numbers," *New York Times Magazine*, 3 January 1999; John Tierney, interview by author, 16 March 2011.

2. Tierney, "Betting on the Planet."

3. Bailey, *Eco-Scam*, 53–54; Bailey, *True State of the Planet*, 2; Bast, Hill, and Rue, *Eco-Sanity*, 124; Stephen Moore, "The Coming Age of Abundance," in Bailey, *True State of the Planet*, 132; Pilzer, *God Wants You to Be Rich*, 48, 54; Huber, *Hard Green*, 14.

4. Colin Campbell, "Who's Trying to Corner Tin?" *New York Times*, 9 February 1982; "Commodities: Tin-Price Crisis Threatens to Cause Rift Between Consuming, Producing Nations," *Wall Street Journal*, 24 February

1982; Thoburn, "Tin Industry Since the Collapse of the International Tin Agreement"; "Tin Prices Plunge to 13-Year Lows," *Wall Street Journal*, 20 March 1986, 48; Thoburn, *Tin in the World Economy*, 173–174.

5. Janice Simpson, "Copper Futures, Producers' Prices Continue to Surge," *Wall Street Journal*, 27 September 1979; H. J. Maidenberg, "Commodities: Copper's New-Found Luster," *New York Times*, 28 January 1980; José Luis Mardones, Enrique Silva and Cristián Martínez, "The Copper and Aluminum Industries: A Review of Structural Changes," *Resources Policy* 11, no. 1 (1985): 14.

6. PRE and John Harte, interview by author, July 15, 2011.

7. Kiel, Matheson, and Golembiewski, "Luck or Skill?"

8. Lazarus, *Making of Environmental Law*, 150–161; Samuel P. Hays, *A History of Environmental Politics Since 1945* (Pittsburgh, PA: University of Pittsburgh Press, 2000), 186–188.

9. George Bush, "Earth Resources and Population—Problems and Directions," *Congressional Record*, 8 July 1970, 23189; Robertson, *Malthusian Moment*, 164; Bill Peterson, "In Boston, Bush Sails into Dukakis: Aboard Harbor Ferry, Republican Blasts 'Neglect' of Environment," *Washington Post*, 2 September 1988; Philip Shabecoff, "Environmental Groups Optimistic About Their Prospects with Bush," *New York Times*, 16 November 1988; Cathleen Decker, "Bush Turns His Attention to Shaping Early Agenda," *Los Angeles Times*, 30 December 1988.

10. "William K. Reilly: Oral History Interview," interview by Dennis Williams, September 1995, http://epa.gov/aboutepa/history/publications/print/reilly.html; William K. Reilly, "Aiming Before We Shoot: The Quiet Revolution in Environmental Policy," 26 September 1990, Speech to the National Press Club, http://epa.gov/aboutepa/history/topics/risk/02.html. Bush also appointed Michael Deland, an outspoken former EPA regional administrator with strong environmental credentials; for Deland and Reilly's account of the Bush administration's environmental commitment and accomplishments, see "Discussant: Michael R. Deland," and "Discussant: William K. Reilly" in Himelfarb and Perotti, *Principle over Politics?* 409–412, 416–420. See also Larry Michlin, "Crisis and Opportunity: American Energy Policy during the Bush Years and Beyond," in Himelfarb and Perotti, *Principle over Politics?* 377–405.

11. Gerald F. Seib, "He's Against Acid Rain; He's Also the Enemy, Say Many Environmentalists; He's John Sununu," *Wall Street Journal*, 2 March 1990; George Bush, "Remarks at an Environmental Agreement Signing Ceremony at the Grand Canyon, Arizona," September 18, 1991, http://www.presidency.ucsb.edu/ws/?pid=19994; Norman J. Vig, "Presidential Leadership and the Environment: From Reagan and Bush to Clinton," in Vig and Kraft, *Environmental Policy in the 1990s*, 83–85.

12. George Bush, "Remarks at an Environmental Agreement Signing

Ceremony at the Grand Canyon, Arizona," September 18, 1991; US Environmental Protection Agency, Office of Air and Radiation, *The Benefits and Costs
of the Clean Air Act from 1990 to 2020: Final Report* (Washington, DC: US
Environmental Protection Agency, 2011), table 2-5; David Malakoff, "Taking
the Sting Out of Acid Rain," *Science* 330 (2010): 910–911; Chan, Stavins,
Stowe, and Sweeney, *SO2 Allowance Trading System and the Clean Air Act
Amendments of 1990.* According to one analysis, the low cost of emission
reductions resulted partly from railroad deregulation and increased use of
low-sulfur coal; see Ellerman and Montero, "Declining Trend in Sulfur
Dioxide Emissions"; and Busse and Keohane, "Market Effects of Environmental Regulation.".

   13. See, e.g., W. J. McG. Tegart, G. W. Sheldon, and D. C. Griffiths, eds.,
Climate Change: The IPCC Impacts Assessment, Report prepared for Inter-
governmental Panel on Climate Change by Working Group II (Canberra:
Australian Government Publishing Service, 1990), 1–5. See also Intergovernmental Panel on Climate Change, *Climate Change, 1995: Impacts, Adaptations
and Mitigation of Climate Change: Scientific-Technical Analyses: Contribution of
WGII to the Second Assessment Report of the Intergovernmental Panel on Climate
Change,* ed. Robert Tony Watson, Marufu C. Zinyowera, and Richard H. Moss
(New York: Cambridge University Press, 1996), ix–x, 3; Schneider, *Science
as a Contact Sport,* 132–154; and Oreskes and Conway, *Merchants of Doubt,*
169–215.

   14. US House of Representatives, *Congressional Record,* 19 May 1992,
11831–11832. The congressmen drew particularly from *Trashing the Planet,* a
book by former Washington governor Dixy Lee Ray, a conservative Democrat
and marine biologist by training. Ray and Guzzo, *Trashing the Planet.*

   15. Julie Hollar, "The Delay Chronicles," *Texas Observer,* 4 February
2000; Peter Perl, "Absolute Truth," *Washington Post,* 13 May 2001: W12;
Congressional Record, 19 May 1992, 11832–11833.

   16. George Bush: "Address to the United Nations Conference on
Environment and Development in Rio de Janeiro, Brazil," June 12, 1992,
http://www.presidency.ucsb.edu/ws/?pid=21075; Martín Carcasson, "Prudence, Procrastination, or Politics: George Bush and the Earth Summit of
1992," in Medhurst, *Rhetorical Presidency of George H. W. Bush,* 119–148; Ann
Devray, "Upbeat President, Charged by Polls, Steps Up Rhetoric," *Washington
Post,* 30 October 1992.

   17. Lazarus, *Making of Environmental Law,* 159; G. B. Trudeau, *The
Bundled Doonesbury: A Pre-Millennial Anthology* (Riverside, NJ: Andrews
McMeel, 2001), 75, 88; Representative Patricia Schroeder, Democrat from
Colorado, entered the *Doonesbury* bid form into the *Congressional Record* to
criticize the Republican Congress for its efforts to "obliterate" the nation's
natural resources. *Congressional Record—Extensions of Remarks,* 23 January
1996, E59.

18. PRE and AHE, *The Population Explosion* (New York: Simon and Schuster, 1990), 9, 13, 174–179; NBC News, "Authors of Healing the Planet Ann and Paul Ehrlich Discuss Global Environmental Crisis," *NBC Today Show*, 28 November 1991; Daily, AHE, and PRE, "Population and Immigration Policy—Paper Number 0053."

19. Meadows et al, *Beyond the Limits*, xv; Barney, *Global 2000 Revisited*. See also Barney, "Global 2000 Report to the President and the Threshold 21 Model," 128.

20. PRE, "Overburdened World," *San Francisco Chronicle*, 29 November 1993; "Protest of Paul Ehrlich's Talk at World Wildlife Fund," 19 December 1995, in Federation for American Immigration Reform Papers, Gelman Archive, George Washington University (hereafter FAIR), box 106, folder "Paul Ehrlich"; Barbara Alexander, executive director, Californians for Population Stabilization to House Subcommittee on Immigration, Refugees, and International Law, 27 September 1989, in FAIR, box 16, folder "Californians for Population Stabilization"; AHE, interview by author, 15 July 2011; Nancy Cleeland, "Sierra Club to Take on Immigration Question," *Los Angeles Times*, 29 September 1997, 1; AHE to Michael K. Dorsey, email communication, 29 May 1997, copy in possession of author; AHE to bod-plus@lists.sierraclub.org, email communication, 16 January 1999, copy in possession of author; Carl Pope, "Ways and Means: Moving On: Lessons of the Immigration Debate," *Sierra Magazine*, July–August 1998; John H. Cushman Jr., "An Uncomfortable Debate Fuels a Sierra Club Election: 2 Views on Immigration Divide the Group," *New York Times*, 5 April 1998; John H. Cushman Jr., "Sierra Club Rejects Move to Oppose Immigration," *New York Times*, 26 April 1998. The issue would surface again with a subsequent effort to seize control of the Sierra Club board. See Kenneth R. Weiss, "Sierra Club Members Vote to Stay Neutral in the Immigration Debate," *Los Angeles Times*, 26 April 2005.

21. Nordhaus, Stavins, and Weitzman, "Lethal Model 2," 5, 7, 37; Thomas Lippman, "Report Warns of Environmental Crisis," *Washington Post*, 14 Apr 1992: A8.

22. JLS, "Earth's Doomsayers Are Wrong," *San Francisco Chronicle*, 12 May 1995.

23. Ibid.

24. Charles Petit, "Two Stanford Professors Offer to Bet Optimistic Economist," *San Francisco Chronicle*, 18 May 1995; PRE and Stephen H. Schneider, "Wagering on Global Environment," *San Francisco Chronicle*, 17 May 1995.

25. Petit, "Two Stanford Professors Offer to Bet Optimistic Economist"; JLS, "May West and East Eventually Meet," 23 May 1995, draft article, JSP, 8282-95-8-22, box 1, unfiled papers.

26. Petit, "Two Stanford Professors Offer to Bet Optimistic Economist";

JLS, "Another Sure Bet on Earth Day," *Wall Street Journal*, 22 April 1997; PRE and Schneider, "Bets and 'Ecofantasies.'"

27. Charles McCoy, "When the Boomster Slams the Doomster, Bet on a New Wager," *Wall Street Journal*, 5 June 1995; David M. Simon, Judith Simon Garrett, and Daniel H. Simon, "Come on Professor, Give Dad a Break," letter to the editor, *Wall Street Journal*, 15 August 1995; JLS, "Truth Is Humanity's Best Hope," *Oakland Tribune*, 24 April 1996; JLS, *Hoodwinking the Nation*; PRE, *Betrayal of Science and Reason*; JLS, "Any Truth in the Balance?" *Washington Times*, 17 February 1992; JLS, "Doing Fine on Planet Earth," *Washington Times*, 21 April 1995; JLS, "Is Economics Dead and Gone?" *Washington Times*, 20 April 1997; William Allen, "Raven, Naysayer Play Point, Counterpoint on Species Extinction," *Saint Louis Post-Dispatch*, 3 November 1993; JLS, "Material Welfare and Standard of Living," *Cato Institute Book Forum*, 27 March 1997, http://www.c-spanvideo.org/program/79990-1; see also Wildavsky, *But Is It True?*

28. JLS, "What the Starvation Lobby Eschews . . . ," *Wall Street Journal*, 18 November 1996; JLS, "Believe It or Not, Energy Is Becoming More Abundant," *Rocky Mountain Gazette Telegraph*, 8 August 1993; JLS, "Another Sure Bet on Earth Day," *Wall Street Journal*, 22 April 1997; "Just as the number of points in a one-inch line can never been counted, even in principle, the quantity of natural resources that might be available to us, and the quantity of services that they can give us, can never be known." JLS, "Earth's Resources: All But Infinite," *Providence Journal*, 2 February 1997.

29. "New Book by Paul and Anne Ehrlich Strikes Back at 'Brownlash,'" news release, Stanford News Service, 11 October 1996, http://news.stanford.edu/pr/96/961021ehrlich.html; Michael Learmonth, "Crowd Control," *Metro-Los Angeles Weekly*, 19–25 February 1998;PRE, "Environmental Anti-Science"; for earlier mockery of economic understandings of basic science, see PRE and AHE, "Ecoscience"; *Mother Earth News*, May–June 1986, 110–111; Norman Myers, "Review: Putting the Brownlashers Straight," *BioScience* 47, no. 3 (1997): 182–183. In 1992, Norman Myers faced off with JLS in a debate at Columbia University, later published as Norman Myers and JLS, *Scarcity or Abundance?* On the meaning of famine, see, e.g., Cullather, *Hungry World*, 205–231; and Amartya Sen, *Poverty and Famines*.

30. "Citation for Anne and Paul Ehrlich," 1995 Heinz Award Brochure, http://www.heinzawards.net/recipients/paul_anne_ehrlich; "Eminent Ecologist Award," *Bulletin of the Ecological Society of America* 83, no. 1 (2002): 17.

31. Ed Regis, "The Doomslayer," *Wired*, February 1997; JLS, *Good Mood*, epilogue. See, e.g., JLS to David Siegmund, 6 January 1995, draft, and "Precis of Efron-Simon Dispute," draft, 9 February 1995, JSP, 8282-95-8-22, box 3, unfiled papers; $72,893 in 1988 dollars equaled $100,180 in 1997, adjusted for inflation. "University of Maryland, Budget with Salary Detail, 1988," 200,

and "University of Maryland, Budget with Salary Detail, 1997," 189, both in Maryland Room, University of Maryland, College Park; JLS, *Life Against the Grain*, 313–314; RJS, interview by author, 22 July 2009.

32. JLS to William Kruskal, 5 November 1986, JSP, 8282-89-08-01, box 4, unfiled papers; JLS, *Life Against the Grain*, 313–314.

33. David M. Simon, correspondence with author, 20 July 2011; JLS, "The Volunteer Auction Plan for Airline Oversales: Saga of a Scheme That Finally Flew," 17 November 1993, unpublished manuscript, JSP, 8282-94-09-01, box 2, unfiled papers; JLS and RJS, letter to Friends, 20 June 1994, draft, JSP, 8282-95-8-22, box 2, unfiled papers.

34. RJS, interview by author, 15 June 2010; Kenneth N. Gilpin, "Julian Simon, 65, Optimistic Economist, Dies," *New York Times*, 12 February 1998.

35. "The Anti-Doomsayer," *Orange County Register*, 11 February 1998; Philip Terzian, "Chicken Little's Enemy," *Providence Journal-Bulletin*, 15 February 1998; Maggie Gallagher, "It Was a Very Bad Bet to Doubt Prophet of Plenty," *New York Post*, 16 February 1998; Thomas Sowell, "Warrior in a 200-Year War," *Tulsa World*, 13 February 1998; Moore, "Julian Simon Remembered"; John Tierney, "Optimism by the Numbers," *New York Times Magazine*, 3 January 1999.

36. Fred L. Smith, "Dedication: Julian Simon—An Appreciation," in Bailey, *Earth Report 2000*, xiv–xvi; Terry L. Anderson later wrote that Simon "built the coffin for neo-Malthusian ideas." Terry L. Anderson, "Environmental Quality," in *The Concise Encyclopedia of Economics*, ed. David R. Henderson (Indianapolis, IN: Liberty Fund, 2007); Barun Mitra, "Sell the Tiger to Save It," *New York Times*, 15 August 2006; Ben Wattenberg, "Malthus, Watch Out," *Wall Street Journal*, 11 February 1998.

37. Lomborg, *Skeptical Environmentalist*, xix, xxiii, 328 (emphasis in original), 330.

38. Ibid., 318, 329, 348; Bjørn Lomborg, "The Environmentalists Are Wrong," *New York Times*, 26 August 2002; Bjørn Lomborg, "Chill Out: Stop Fighting over Global Warming," *Washington Post*, 7 October 2007; Bjørn Lomborg, "A Better Way Than Cap and Trade," *Washington Post*, 26 June 2008.

39. See, e.g., "Doomsday Postponed; Environmental Scrutiny," *Economist*, 8 September 2001, 89–90; E. O. Wilson, "On Bjørn Lomborg and Extinction," *Grist*, 12 December 2001, http://grist.org/article/point/; Stuart Pimm and Jeff Harvey, "No Need to Worry About the Future," *Nature* 414 (2001): 149–150; Friel, *Lomborg Deception;* Harrison, "Peer Review, Politics and Pluralism"; Paul R. Ehrlich, email communication with author, 16 January 2013; Andrew Revkin, "Environment and Science: Danes Rebuke a 'Skeptic,'" *New York Times*, 8 January 2003; Andrew Revkin, "Danish Ethics Panel Censured for Critique of Book," *New York Times*, 23 December 2002; "Bjorn Lomborg Cream Pied by Mark Lynas," 4 September 2001, http://www.youtube.com/watch?v=TOg8IqkS4PA.

40. Roger A. Pielke Jr. and Steve Rayner, "Editors' Introduction," *Environmental Science and Policy* 7, no. 5 (2004): 255. See also Herrick, "Objectivity Versus Narrative Coherence"; Oreskes, "Science and Public Policy," 369, 376, 381.

41. Richard Benedetto, "Cheney's Energy Plan Focuses on Production," *USA Today*, 1 May 2001. For an insider account that emphasizes the Bush administration's efforts, in a partisan political climate, to balance costs and benefits in achieving environmental improvements, see John D. Graham, *Bush on the Home Front: Domestic Policy Triumphs and Failures* (Bloomington: Indiana University Press, 2010); for the counterview on Bush environmental policy, see Kennedy, *Crimes Against Nature;* Mooney, *Republican War on Science;* see also Meg Jacobs, "Wreaking Havoc From Within: George W. Bush's Energy Policy in Historical Perspective," in *The Presidency of George W. Bush: A First Historical Assessment*, ed. Julian E. Zelizer (Princeton, NJ: Princeton University Press, 2010), 139–168. Graham outlines his ideas about balancing tradeoffs in regulatory policy in Graham and Wiener, *Risk Versus Risk.*

Chapter Six: Betting the Future of the Planet

1. For an analysis of global health trends since 1970, see Wang et al., "Age-Specific and Sex-Specific Mortality in 187 Countries, 1970–2010"; for a compelling and brief visual representation of global trends in health and wealth, see Hans Rosling, "200 Countries, 200 Years, 4 Minutes," http://www .youtube.com/watch?v=jbkSRLYSojo. The statistics underlying Rosling's work are available at www.gapminder.org.

2. Daily, AHE, and PRE, "Optimum Human Population Size," *Population and Environment*, 469, 474. The Ehrlichs repeated this number in PRE and AHE, "Population, Development and Human Natures," 161. John Holdren denied that his or PRE's statements about the future were ever predictions, arguing instead that his statements were "intended not as predictions but as projections." "Nominations to the Executive Office of the President and the Department of Commerce: Hearing before the Committee on Commerce, Science, and Transportation," 12 February 2009, 111th Cong., 1st sess. (Washington, DC: GPO, 2010), 105. More generally, John Holdren has argued that correctness of predictions is a poor measure of their usefulness: "To put too much emphasis on the correctness or incorrectness of particular predictions, however, is to miss the main point of writing usefully about the future. The idea is not to be 'right,' but to illuminate the possibilities in a way that both stimulates sensible debate about the sort of future we want and facilitates sound decisions about getting from here to there." John Holdren, "Introduction," in Smith, Fesharaki, and Holdren, *Earth and the Human Future*, 79. Ehrlich described provocative, pessimistic predictions as a strategy in 1996: "It would be intellectually satisfying to say the real impact is

through reasoned discourse. But in my view the real impact isn't in reasoned discourse. Media attention, press coverage and, if necessary, alarmism at least set an agenda. And that way you can have a debate." Michael Wines, "The Sky Is Falling: Three Cheers for Chicken Little," *New York Times*, 29 December 1996.

3. In 2005, Houston investment banker Matthew Simmons published *Twilight in the Desert*, a dour book that said that declining Saudi oil production would soon cause a major global economic shock. In an interview with the *New York Times Magazine*, Simmons predicted that oil prices soon would soar into the triple digits—and not the "low triple digits." Skeptical of Simmons's forecast, the opinion columnist John Tierney, friend and promoter of JLS, called Simmons to see if the investor would put money behind his forecasts. Simmons agreed to bet Tierney $5,000 that the price of oil, adjusted for inflation, would average more than $200 per barrel over the year 2010. At the time of their bet in 2005, oil prices were around $65. After Tierney told RJS about the wager, she joined Tierney's side of the bet, staking $2,500. Matthew Simmons died in August 2010, so he did not live to see the conclusion of his bet with Tierney. But even at the time of his death, the outcome of the Simmons-Tierney wager was clear. Oil prices did rise from 2005 to 2010. But they never approached Simmons's predicted $200 average, even for a single day, let alone over the entire year. From a price per barrel of $65 at the time of their bet in 2005, oil prices climbed to a high of $145 in the summer of 2008 before falling back sharply. For the year 2010, oil prices averaged just $80, or $71 in inflation-adjusted 2005 dollars, only a little higher than the price at the time of their bet. Simmons's colleagues paid up with Tierney and RJS, and Tierney crowed about his victory in a column titled "Economic Optimism? Yes, I'll Take That Bet." Peter Maass, "The Breaking Point," *New York Times Magazine*, 21 August 2005; Simmons, *Twilight in the Desert*. John Tierney, "10,000 Dollar Question," *New York Times*, 23 August 2005; John Tierney, "Economic Optimism? Yes, I'll Take That Bet," *New York Times*, 27 December 2010. For the falling price of polysilicon, see Saqib Rahim and Peter Behr, "How Well Did DOE Know Solyndra's Technology—and Its Market Vulnerabilities?" *Climatewire*, 15 September 2011.

4. PRE and John Harte, interview by author, 15 July 2011; PRE and AHE, "Can a Collapse of Global Civilization Be Avoided?" *Proceedings of the Royal Society B* 280 (January 2013): 2012–2845; Goodstein, Out of Gas, 15; Kunstler, *Long Emergency*, 20; James Howard Kunstler, "Foreword," in McCommons, *Waiting on a Train*, xi. For the use of the Easter Islands story as a parable about the possible collapse of civilization, see Diamond, *Collapse*, as well as Turner, "Vindication of a Public Scholar," 56; for a comprehensive critique of Diamond's interpretation, see McAnany and Yoffee, *Questioning Collapse*. For other warnings and discussion of the economic threat of peak

oil and resource scarcity, see Jeremy Grantham, "Time to Wake Up: Days of Abundant Resources and Falling Prices Are Over Forever," GMO Quarterly Letter, April 2011; Worth, *Peak Oil and the Second Great Depression*, xv; Heinberg, *End of Growth* (Gabriola Island, BC: New Society, 2011), 1–2; Heinberg, *Peak Everything;* Michael C. Ruppert, *Confronting Collapse: The Crisis of Energy and Money in a Post Peak Oil World* (White River Junction, VT: Chelsea Green, 2009); see also Klare, *Race for What's Left;* James Elser and Stuart White, "Peak Phosphorus," *Foreign Policy,* 20 April 2010, http:// www.foreignpolicy.com/articles/2010/04/20/peak_phosphorus; James D. Hamilton, "Causes and Consequences of the Oil Shock of 2007–08."

5. William D. Nordhaus, "The Ecology of Markets," *Proceedings of the National Academy of Sciences* 89, no. 3 (1992): 843–850; Akerlof and Shiller, *Animal Spirits;* Sabin, *Crude Politics;* White, *Railroaded;* Lewis, *Big Short;* N. Gregory Mankiw, "One Answer to Global Warming: A New Tax," *New York Times,* 16 September 2007; N. Gregory Mankiw, "Raise the Gas Tax," *Wall Street Journal,* 20 October 2006.

6. Posner, *Public Intellectuals,* 134; Robert Bradley Jr., "Climategate Did Not Begin with Climate (Remembering Julian Simon and the Storied Intolerance of Neo-Malthusians)," 8 December 2009, http://www.master resource.org/2009/12/climategate-did-not-begin-with-climate-remembering -julian-simon-and-the-intolerance-of-neo-malthusianism/; Steven F. Hayward, "All the Leaves Are Brown," *Claremont Review of Books* 9, no. 1 (2008), http://www.claremont.org/publications/crb/id.1588/article_detail .asp; George Will, "Apocalypse Not: Ingenuity Thwarts Doomsday," *Hartford Courant,* 17 August 2012; Fred Smith, "Foreword," in Robert Bradley Jr., *Julian Simon and the Triumph of Energy Sustainability* (Washington, DC: American Legislative Exchange Council, 2000), 13. See also Anderson and Huggins, *Greener Than Thou,* 57 ("gloom-and-doom provides a pulpit for greener-than-thou regulatory environmentalism").

7. James Inhofe, "The Science of Climate Change: Senate Floor Statement," in Bill McKibben, ed., *The Global Warming Reader: A Century of Writing About Climate Change* (New York: Penguin, 2012), 185; Aaron Wildavsky, "Introduction," in Robert C. Balling, *The Heated Debate: Greenhouse Predictions Versus Climate Reality* (San Francisco: Pacific Research Institute for Public Policy, 1992), 1; Rush Limbaugh, "Environmentalist Wacko: Climate Change Skeptics Are Sick," *The Rush Limbaugh Show,* 2 April 2012, http://www.rushlimbaugh.com/daily/2012/04/02/environmentalist _wacko_climate_change_skeptics_are_sick; Coll, *Private Empire,* 84–89; Oreskes and Conway, *Merchants of Doubt;* Pooley, *Climate War.*

8. See, e.g., James Lovelock, *The Revenge of Gaia: Earth's Climate Crisis and the Fate of Humanity* (New York: Basic Books, 2007), xiv; for Lovelock's disavowal of this earlier prediction, see Ian Johnston, "'Gaia' Scientist James

Lovelock: I Was 'Alarmist' About Climate Change," msnbc.com, 23 April 2012, http://worldnews.nbcnews.com/_news/2012/04/23/11144098-gaia -scientist-james-lovelock-i-was-alarmist-about-climate-change; for a survey of studies of the economic impact of climate change, see Richard S. J. Tol, "The Economic Effects of Climate Change," *Journal of Economic Perspectives* 23, no. 2 (2009): 29–51.

Select Bibliography

Archival Collections

American Academy for the Advancement of Science Papers, Washington, DC
American Heritage Center, University of Wyoming
    Harold J. Barnett Papers
    Julian Lincoln Simon Papers
    James G. Watt Papers
    Paul Weyrich Papers
Dartmouth College, Rauner Special Collections Library
    Donella Meadows Papers
George Washington University, Gelman Library
    Federation for American Immigration Reform Papers
Hoover Institution Archives, Stanford University
    John Armand Busterud Papers
    Milton Friedman Papers
    Gottfried Haberler Papers
    Friedrich A. von Hayek Papers
    Fritz Machlup Papers
Library of Congress
    Johnny Carson Papers
National Archives
    Jimmy Carter Library
Stanford University, Special Collections and University Archives
    Paul and Anne Ehrlich Papers
University of California–Berkeley, Bancroft Library
    Sierra Club Papers

University of Illinois Archives
    Julian L. Simon Papers
    Rita J. Simon Papers
Yale University Library, Manuscripts and Archives
    G. Evelyn Hutchinson Papers

Published Sources

Abernethy, Virginia. "How Julian Simon Could Win the Bet and Still Be Wrong." *Population and Environment* 13, no. 1 (1991): 3–7.

Ahlburg, Dennis A. "Julian Simon and the Population Growth Debate." *Population and Development Review* 24, no. 2 (1998): 317–327.

Akerlof, George A., and Robert J. Shiller. *Animal Spirits: How Human Psychology Drives the Economy, and Why It Matters for Global Capitalism.* Princeton, NJ: Princeton University Press, 2009.

Akins, James. "The Oil Crisis: This Time the Wolf Is Here." *Foreign Affairs* 51 (April 1973): 462–490.

Aligica, Paul Dragos. *Prophecies of Doom and Scenarios of Progress: Herman Kahn, Julian Simon, and the Prospective Imagination.* New York: Continuum International, 2007.

Altenberg, Lee. "From the Bedroom to the Bomb: An Interview with Paul Ehrlich." *Stanford Daily,* April 1, 1983.

Anderson, Terry L., and Laura E. Huggins. *Greener Than Thou.* Stanford, CA: Hoover Institution Press, 2008.

Anderson, Terry L., and Donald Leal. *Free Market Environmentalism.* Rev. ed. New York: Palgrave, 2001.

Bailey, Ron. *Eco-Scam: The False Prophets of Ecological Apocalypse.* New York: St. Martin's Press, 1993.

Bailey, Ronald, ed. *Earth Report 2000: Revisiting the True State of the Planet.* New York: McGraw-Hill, 2000.

———. *Global Warming and Other Eco-Myths: How the Environmental Movement Uses False Science to Scare Us to Death.* Roseville, CA: Prima, 2002.

———. *The True State of the Planet: Ten of the World's Premier Environmental Researchers in a Major Challenge to the Environmental Movement.* New York: Free Press, 1995.

Baker, Jeffrey J. W. "Science, Birth Control, and the Roman Catholic Church." *BioScience* 20, no. 3 (1970): 143–151.

Balint, Peter J. "How Ethics Shape the Policy Preferences of Environmental Scientists: What We Can Learn from Lomborg and His Critics." *Politics and the Life Sciences* 22, no. 1 (2003): 14–23.

Barlow, Nora, ed. *The Autobiography of Charles Darwin, 1809–1882; With the Original Omissions Restored.* London: Collins, 1958.

Barnett, Harold J., and Chandler Morse. *Scarcity and Growth: The Economics of Natural Resource Availability.* Washington, DC: Resources for the Future, 1963.

Barney, Gerald O. "Global 2000 Report to the President and the Threshold 21 Model: Influences of Dana Meadows and System Dynamics." *System Dynamics Review* 18, no. 2 (2002): 123–136.

———. *Global 2000 Revisited: What Shall We Do?* Arlington, VA: Millennium Institute, 1993.

Barney, Gerald O., ed. *The Unfinished Agenda: The Citizen's Policy Guide to Environmental Issues; A Task Force Report Sponsored by the Rockefeller Brothers Fund.* New York: Crowell, 1977.

Bast, Joseph L., Peter Jensen Hill, and Richard Rue. *Eco-Sanity: A Common Sense Guide to Environmentalism.* Lanham, MD: Madison Books, 1994.

Baumol, William J. "On Taxation and the Control of Externalities." *American Economic Review* 62, no. 3 (1972): 307–322.

Bender, Thomas, and Carl E. Schorske, eds. *American Academic Culture in Transformation: Fifty Years, Four Disciplines.* Princeton, NJ: Princeton University Press, 1998.

Bentham, Jeremy. *A Fragment on Government; or, A Comment on the Commentaries . . . .* London: E. Wilson and W. Pickering, 1823.

———. *An Introduction to the Principles of Morals and Legislation.* London: W. Pickering and R. Wilson, 1823.

Bernstein, Michael A. *A Perilous Progress: Economists and Public Purpose in Twentieth-Century America.* Princeton, NJ: Princeton University Press, 2001.

Berry, Wendell. *A Continuous Harmony: Essays Cultural and Agricultural.* New York: Harcourt Brace Jovanovich, 1972.

Boserup, Ester. *The Conditions of Agricultural Growth: The Economics of Agrarian Change Under Population Pressure.* Chicago: Aldine, 1965.

Bourne, Peter G. *Jimmy Carter: A Comprehensive Biography from Plains to Post-Presidency.* New York: Scribner, 1997.

Brinkley, Douglas. "Bringing the Green Revolution to Africa: Jimmy Carter, Norman Borlaug, and the Global 2000 Campaign." *World Policy Journal* 13, no. 1 (1996): 53–62.

———. *The Unfinished Presidency: Jimmy Carter's Journey Beyond the White House.* New York: Viking, 1998.

Brookhaven National Laboratory, Biology Department. *Diversity and Stability in Ecological Systems; Report of Symposium Held May 26–28, 1969.* Brookhaven Symposia in Biology, no. 22. Upton, NY: Brookhaven National Laboratory, 1969.

Brown, Harrison. *The Challenge of Man's Future: An Inquiry Concerning the Condition of Man During the Years That Lie Ahead.* New York: Viking, 1954.

Brownlee, W. Elliot, and Hugh Davis Graham, eds. *The Reagan Presidency: Pragmatic Conservatism and Its Legacies*. Lawrence: University Press of Kansas, 2003.

Buckley, James Lane. *Freedom at Risk: Reflections on Politics, Liberty, and the State*. New York: Encounter Books, 2010.

Burford, Anne M., with John Greenya. *Are You Tough Enough?* New York: McGraw-Hill, 1986.

Burrough, Bryan. *The Big Rich: The Rise and Fall of the Greatest Texas Oil Fortunes*. New York: Penguin, 2009.

Busse, Meghan R., and Nathaniel O. Keohane. "Market Effects of Environmental Regulation: Coal, Railroads, and the 1990 Clean Air Act." *RAND Journal of Economics* 38, no. 4 (2007): 1159–1179.

Cannon, Lou. *Governor Reagan: His Rise to Power*. New York: PublicAffairs, 2003.

Carson, Rachel. *Silent Spring*. Boston: Houghton Mifflin, 1962.

Carter, Jimmy. *Living Faith*. New York: Times Books, 1996.

———. *A Remarkable Mother*. New York: Simon and Schuster, 2008.

———. *Why Not the Best?* Nashville, TN: Broadman, 1975.

Carter, Lillian, and Gloria Carter Spann. *Away from Home: Letters to My Family*. New York: Simon and Schuster, 1977.

Carter, Luther J. "The Population Crisis: Rising Concern at Home." *Science* 166 (1969): 722–726.

Carter, Rosalynn. *First Lady from Plains*. Boston: Houghton Mifflin, 1984.

Cawley, R. McGreggor. *Federal Land, Western Anger: The Sagebrush Rebellion and Environmental Politics*. Lawrence: University Press of Kansas, 1993.

Chan, Gabriel, Robert Stavins, Robert Stowe, and Richard Sweeney. *The SO2 Allowance Trading System and the Clean Air Act Amendments of 1990: Reflections on Twenty Years of Policy Innovation*. NBER Working Paper no. 17845. Cambridge, MA: National Bureau of Economic Research, 2012.

Clarke, Robin. *London Under Attack: The Report of the Greater London Area War Risk Study Commission*. Oxford: Basil Blackwell, 1986.

Coates, Peter A. *The Trans-Alaska Pipeline Controversy: Technology, Conservation, and the Frontier*. Bethlehem, PA: Lehigh University Press, 1991.

Cohen, Joel E. *How Many People Can the Earth Support?* New York: W. W. Norton, 1995.

Cole, H. S. D., Christopher Freeman, Marie Jahoda, and K. L. R. Pavitt, eds. *Models of Doom: A Critique of "The Limits to Growth."* New York: Universe Books, 1973.

Coll, Steve. *Private Empire: ExxonMobil and American Power*. New York: Penguin, 2012.

Commoner, Barry. *The Closing Circle: Nature, Man, and Technology*. New York: Knopf, 1971.

———. *Science and Survival.* New York: Viking, 1966.

Connelly, Matthew James. *Fatal Misconception: The Struggle to Control World Population.* Cambridge, MA: Belknap Press of Harvard University Press, 2008.

Costanza, Robert, et al. *An Introduction to Ecological Economics.* Boca Raton, FL: St. Lucie Press / International Society for Ecological Economics, 1997.

Critchlow, Donald T. *The Conservative Ascendancy: How the GOP Right Made Political History.* Cambridge, MA: Harvard University Press, 2007.

———. *Intended Consequences: Birth Control, Abortion, and the Federal Government in Modern America.* New York: Oxford University Press, 1999.

Cronon, William, ed. *Uncommon Ground: Rethinking the Human Place in Nature.* New York: W. W. Norton, 1996.

Cullather, Nick. *The Hungry World: America's Cold War Battle Against Poverty in Asia.* Cambridge, MA: Harvard University Press, 2010.

Cunningham, Barbara, ed. *The New Jersey Ethnic Experience.* Union City, NJ: W. H. Wise, 1977.

Daily, Gretchen C., Anne H. Ehrlich, and Paul R. Ehrlich. "Optimum Human Population Size." *Population and Environment* 15, no. 6 (1994): 469–475.

———. "Population and Immigration Policy—Paper Number 0053." Unpublished paper, 1994.

Daly, Herman E. "On Economics as a Life Science." *Journal of Political Economy* 76 (1968): 392–406.

———. *Steady State Economics.* San Francisco: W. H. Freeman, 1977.

Dasgupta, Partha, and Geoffrey Heal. "The Optimal Depletion of Exhaustible Resources." *Review of Economic Studies* 41 (1974): 3–28.

Davis, Frederick R. *The Man Who Saved Sea Turtles.* Oxford: Oxford University Press, 2007.

Davis, Joseph S. "Population and Resources." *Journal of the American Statistical Association* 45 (1950): 346–349.

De Bell, Garrett. *The Environmental Handbook.* New York: Ballantine, 1970.

Deffeyes, Kenneth S. *Hubbert's Peak: The Impending World Oil Shortage.* Rev. ed. Princeton, NJ: Princeton University Press, 2003.

DeMuth, Christopher. "OIRA at Thirty." *Administrative Law Review* 63, special ed. (2011): 15–25.

Dewey, Scott Hamilton. *Don't Breathe the Air: Air Pollution and U.S. Environmental Politics, 1945–1970.* College Station: Texas A&M University Press, 2000.

Diamandis, Peter H., and Steven Kotler. *Abundance: The Future Is Better Than You Think.* New York: Free Press, 2012.

Diamond, Jared M. *Collapse: How Societies Choose to Fail or Succeed.* New York: Viking, 2005.

———. *Guns, Germs, and Steel: The Fates of Human Societies.* New York: W. W. Norton, 1997.
Douglas, James. "Paul Ehrlich: An Interview." *Metamorphoses,* Winter 1988–1989, 31.
Douglas, Mary, and Aaron B. Wildavsky. *Risk and Culture: An Essay on the Selection of Technical and Environmental Dangers.* Berkeley: University of California Press, 1982.
Doyle, Jack. *Taken for a Ride: Detroit's Big Three and the Politics of Pollution.* New York: Four Walls Eight Windows, 2000.
Drew, Elizabeth. *Portrait of an Election: The 1980 Presidential Campaign.* London: Routledge and Kegan Paul, 1981.
Dunlap, Riley E., and Angela G. Mertig. *American Environmentalism: The U.S. Environmental Movement, 1970–1990.* New York: Taylor and Francis, 1992.
Dunlap, Thomas R. *Faith in Nature: Environmentalism as Religious Quest.* Seattle: University of Washington Press, 2004.
Easterbrook, Gregg. *A Moment on the Earth: The Coming Age of Environmental Optimism.* New York: Viking, 1995.
———. *The Progress Paradox: How Life Gets Better While People Feel Worse.* New York: Random House, 2003.
Easterlin, Richard A. "Economic-Demographic Interactions and Long Swings in Economic Growth." *American Economic Review* 56, no. 5 (1966): 1063–1104.
———. "Effects of Population Growth on the Economic Development of Developing Countries." *Annals of the American Academy of Political and Social Science* 369 (1967): 98–108.
Easton, Robert Olney. *Black Tide: The Santa Barbara Oil Spill and Its Consequences.* New York: Delacorte, 1972.
Edgerton, F. N., ed. *History of American Ecology.* New York: Arno, 1977.
Efron, Edith. *The Apocalyptics: Cancer and the Big Lie.* New York: Simon and Schuster, 1984.
Egan, Michael. *Barry Commoner and the Science of Survival: The Remaking of American Environmentalism.* Cambridge, MA: MIT Press, 2007.
Ehrlich, Anne H. "The Human Population—Size and Dynamics." *American Zoologist* 25, no. 2 (1985): 395–406.
Ehrlich, Anne H., and John W. Birks, eds. *Hidden Dangers: Environmental Consequences of Preparing for War.* San Francisco: Sierra Club Books, 1990.
Ehrlich, Anne H., and Paul R. Ehrlich. *Earth.* New York: F. Watts, 1987.
Ehrlich, Paul R. "AIBS News." *BioScience* 37 (1987): 757–763.
———. "Cheap Nuclear Power Could Lead to Civilization's End." *Los Angeles Times,* 27 May 1975.
———. "An Economist in Wonderland." *Social Science Quarterly* 62, no. 1 (1981): 44–49.

———. "Environmental Anti-Science." *TREE* 11 (September 1996): 393.

———. "Environmental Disruption: Implications for the Social Sciences." *Social Science Quarterly* 62, no. 1 (1981): 7–29.

———. *How to Know the Butterflies.* Dubuque, IA: W. C. Brown, 1961.

———. *Human Natures: Genes, Cultures, and the Human Prospect.* Washington, DC: Shearwater Books, 2000.

———. *The Machinery of Nature.* New York: Simon and Schuster, 1986.

———. "The Population Biology of the Butterfly, Euphydryas Editha. II. The Structure of the Jasper Ridge Colony." *Evolution* 19, no. 3 (1965): 327–336.

———. *The Population Bomb.* New York: Ballantine, 1968.

———. "Population Control or Hobson's Choice." In *The Optimum Population for Britain,* ed. L. R. Taylor, 151–162. London: Academic Press, 1970.

———. "Why the Club of Earth." *TREE* 2 (May 1987): 133–135.

———. *A World of Wounds: Ecologists and the Human Dilemma.* Oldendorf/Luhe, Germany: Ecology Institute, 1997.

Ehrlich, Paul R., and L. C. Birch. "The 'Balance of Nature' and 'Population Control.'" *American Naturalist* 101 (March–April 1967): 97–107.

Ehrlich, Paul R., and Gretchen C. Daily. "Population Extinction and Saving Biodiversity." *Ambio* 22, no. 2/3 (1993): 64–68.

Ehrlich, Paul R., and Anne H. Ehrlich. *Betrayal of Science and Reason: How Anti-Environmental Rhetoric Threatens Our Future.* Washington, DC: Island Press, 1996.

———. "Can a Collapse of Global Civilization Be Avoided?" Proceedings of the Royal Society B 280: 2012–2845 (January 2013).

———. *The Dominant Animal: Human Evolution and the Environment.* Washington, DC: Island Press, 2008.

———. "Ecoscience: Nature, Population, and Economics." *Mother Earth News,* May–June 1986, 110–111.

———. *The End of Affluence: A Blueprint for Your Future.* New York: Ballantine, 1974.

———. *Extinction: The Causes and Consequences of the Disappearance of Species.* New York: Random House, 1981.

———. *One with Nineveh: Politics, Consumption, and the Human Future.* Washington, DC: Island Press, 2004.

———. "The Population Bomb Revisited." *Electronic Journal of Sustainable Development* 1, no. 3 (2009): 63.

———. "Population, Development and Human Natures." *Environment and Development Economics* 7 (2002): 158–170.

———. *The Population Explosion.* New York: Simon and Schuster, 1990.

———. *Population Resources Environment: Issues in Human Ecology.* San Francisco: W. H. Freeman, 1970.

———. "The Value of Biodiversity." *Ambio* 21, no. 3 (1992): 219–226.

Ehrlich, Paul R., and Shirley Feldman. *The Race Bomb: Skin Color, Prejudice, and Intelligence.* New York: Quadrangle, 1977.

Ehrlich, Paul R., and Richard L. Harriman. *How to Be a Survivor.* New York: Ballantine, 1971.

Ehrlich, Paul R. and John P. Holdren. "Impact of Population Growth." *Science* 171 (1971): 1212–1217.

———. "Population and Panaceas: A Technological Perspective." *BioScience* 19 (1969): 1065–1071.

———. "Starvation as a Policy." *Saturday Review,* December 4, 1971.

Ehrlich, Paul R., and John P. Holdren, eds. *The Cassandra Conference: Resources and the Human Predicament.* College Station: Texas A&M University Press, 1988.

Ehrlich, Paul R., and Richard W. Holm. "Patterns and Populations." *Science* 137 (1962): 652–657.

———. *The Process of Evolution.* New York: McGraw-Hill, 1963.

Ehrlich, Paul R., and Larry G. Mason. "The Population Biology of the Butterfly Euphydryas Editha. III. Selection and the Phenetics of the Jasper Ridge Colony." *Evolution* 20, no. 2 (1966): 165–173.

Ehrlich, Paul R., and Dennis D. Murphy. "Conservation Lessons from Long-Term Studies of Checkerspot Butterflies." *Conservation Biology* 1, no. 2 (1987): 122–131.

Ehrlich, Paul R., and Robert E. Ornstein. *Humanity on a Tightrope: Thoughts on Empathy, Family, and Big Changes for a Viable Future.* Lanham, MD: Rowman and Littlefield, 2010.

Ehrlich, Paul R., and Peter H. Raven. "Butterflies and Plants: A Study in Coevolution." *Evolution* 18 (1965): 586–608.

Ehrlich, Paul R., and S. H. Schneider. "Bets and 'Ecofantasies.'" *Environmental Awareness* 18 (1995): 47–50.

Ehrlich, Paul R., Loy Bilderback, and Anne H. Ehrlich. *The Golden Door: International Migration, Mexico, and the United States.* New York: Ballantine, 1979.

Ehrlich, Paul R., Anne H. Ehrlich, and Gretchen C. Daily. *The Stork and the Plow: The Equity Answer to the Human Dilemma.* New York: Putnam's, 1995.

Ehrlich, Paul R., Anne H. Ehrlich, and John P. Holdren. *Ecoscience: Population, Resources, Environment.* San Francisco: W. H. Freeman, 1977.

———. *Human Ecology: Problems and Solutions.* San Francisco: W. H. Freeman, 1973.

Ehrlich, Paul R., Richard W. Holm, and Irene L. Brown. *Biology and Society.* New York: McGraw-Hill, 1976.

Ehrlich, Paul R., John P. Holdren, and Richard W. Holm, eds. *Man and the Ecosphere: Readings from "Scientific American."* San Francisco: W. H. Freeman, 1971.

Ehrlich, Paul R., Carl Sagan, Donald Kennedy, and Walter Orr Roberts. *The*

Cold and the Dark: The World After Nuclear War. New York: W. W. Norton, 1984.

Ehrlich, Paul R., et al. "Checkerspot Butterflies: A Historical Perspective." *Science* 188 (1975): 221–228.

———. "Extinction, Reduction, Stability and Increase: The Responses of Checkerspot Butterfly (Euphydryas) Populations to the California Drought." *Oecologia* 46, no. 1 (1980): 101–105.

———. "Long-Term Biological Consequences of Nuclear War." *Science* 222 (1983): 1293–1300.

———. "Weather and the 'Regulation' of Subalpine Populations." *Ecology* 53, no. 2 (1972): 243–247.

Eizenstat, Stuart. Interview. Miller Center, University of Virginia, Jimmy Carter Presidential Oral History Project, January 29–30, 1982.

Ellerman, A. Denny, and Juan-Pablo Montero. "The Declining Trend in Sulfur Dioxide Emissions: Implications for Allowance Prices." *Journal of Environmental Economics and Management* 36, no. 1 (1998): 26–45.

Esty, Daniel C. *Green to Gold: How Smart Companies Use Environmental Strategy to Innovate, Create Value, and Build Competitive Advantage.* New Haven: Yale University Press, 2006.

Ferguson, Thomas, and Joel Rogers, eds. *The Hidden Election: Politics and Economics in the 1980 Presidential Campaign.* New York: Pantheon Books, 1981.

Fiege, Mark. *Republic of Nature.* Seattle: University of Washington Press, 2012.

Fink, Gary M., and Hugh Davis Graham, eds. *The Carter Presidency: Policy Choices in the Post-New Deal Era.* Lawrence: University Press of Kansas, 1998.

Flippen, J. Brooks. *Conservative Conservationist: Russell E. Train and the Emergence of American Environmentalism.* Baton Rouge: Louisiana State University Press, 2006.

———. *Nixon and the Environment.* Albuquerque: University of New Mexico Press, 2000.

Fogel, Robert William. *The Escape from Hunger and Premature Death, 1700–2100: Europe, America and the Third World.* Cambridge: Cambridge University Press, 2004.

Forrester, Jay Wright. *World Dynamics.* Cambridge, MA: Wright-Allen, 1971.

Friedman, Milton. *Capitalism and Freedom.* Chicago: University of Chicago Press, 1962.

Friedman, Milton, and Simon Smith Kuznets. *Income from Independent Professional Practice.* New York: National Bureau of Economic Research, 1945.

Friedman, Milton, and George Joseph Stigler. *Roofs or Ceilings? The Current Housing Problem.* Irvington-on-Hudson, NY: Foundation for Economic Education, 1946.

Friel, Howard. *The Lomborg Deception: Setting the Record Straight About Global Warming*. New Haven: Yale University Press, 2010.

Fussler, Herman Howe, and Julian L. Simon. *Patterns in the Use of Books in Large Research Libraries*. Chicago: University of Chicago Press, 1969.

Gardner, Dan. *Future Babble: Why Expert Predictions Are Next to Worthless, and You Can Do Better*. New York: Dutton, 2011.

Gelbspan, Ross. *The Heat Is On: The High Stakes Battle over Earth's Threatened Climate*. Reading, MA: Addison-Wesley, 1997.

Gilmour, Robert S., and John A. McCauley. "Environmental Preservation and Politics: The Significance of 'Everglades Jetport.'" *Political Science Quarterly* 90 (1975–76): 719–738.

Glad, Betty. *Jimmy Carter: In Search of the Great White House*. New York: W. W. Norton, 1980.

Glenn, Brian J., and Steven Michael Teles, eds. *Conservatism and American Political Development*. New York: Oxford University Press, 2009.

Glover, Donald R., and Julian L. Simon. "The Effect of Population Density on Infrastructure: The Case of Road Building." *Economic Development and Cultural Change* 23, no. 3 (1975): 453–468.

Godbold, E. Stanley. *Jimmy and Rosalynn Carter: The Georgia Years, 1924–1974*. New York: Oxford University Press, 2010.

Godwin, William. *Of Population: An Enquiry Concerning the Power of Increase in the Numbers of Mankind, Being an Answer to Mr. Malthus's Essay on that Subject*. London: Longman, Hurst, Rees, Orme and Brown, 1820.

Goeller, H. E., and Alvin M. Weinberg. "The Age of Substitutability." *Science* 191 (1976): 683–689.

Goldsmith, Edward, et al. *Blueprint for Survival*. Boston: Houghton Mifflin, 1972.

Golley, Frank B. *A History of the Ecosystem Concept in Ecology: More Than the Sum of the Parts*. New Haven: Yale University Press, 1993.

Goodstein, David. *Out of Gas: The End of the Age of Oil*. New York: W. W. Norton, 2005.

Gore, Albert. *The Assault on Reason*. New York: Penguin, 2007.

Gottlieb, Robert. *Forcing the Spring: The Transformation of the American Environmental Movement*. Washington, DC: Island Press, 2005.

Graetz, Michael J. *The End of Energy: The Unmaking of America's Environment, Security, and Independence*. Cambridge, MA: MIT Press, 2011.

Graham, John D. *Bush on the Home Front: Domestic Policy Triumphs and Failures*. Bloomington: Indiana University Press, 2010.

Graham, John D., and Jonathan Baert Wiener, eds. *Risk Versus Risk: Tradeoffs in Protecting Health and the Environment*. Cambridge, MA: Harvard University Press, 1995.

Greenya, John, and Anne Urban. *The Real David Stockman*. New York: St. Martin's, 1986.

Greider, William. "The Education of David Stockman." *Atlantic Monthly,* December 1981.

Halpern, Charles. *Making Waves and Riding the Currents: Activism and the Practice of Wisdom.* San Francisco: Berrett-Koehler, 2008.

Hamilton, James D. "Causes and Consequences of the Oil Shock of 2007–08." *Brookings Paper on Economic Activity,* Spring 2009, 215–283.

Hardin, Garrett. "Multiple Paths to Population Control." *Family Planning Perspectives* 2, no. 3 (1970): 24–26.

———. *Population, Evolution, and Birth Control: A Collage of Controversial Ideas.* 2nd ed. San Francisco: W. H. Freeman, 1969.

———. "The Tragedy of the Commons." *Science* 162 (1968): 1243–1248.

Harrison, Chris. "Peer Review, Politics and Pluralism." *Environmental Science and Policy* 7, no. 5 (2004): 357–368.

Harte, John. *Consider a Spherical Cow: A Course in Environmental Problem Solving.* Mill Valley, CA: University Science Books, 1988.

———. *The Green Fuse: An Ecological Odyssey.* Berkeley: University of California Press, 1993.

Harte, John, and Robert H. Socolow, eds. *The Patient Earth.* New York: Holt, Rinehart and Winston, 1971.

Hawken, Paul, Amory B. Lovins, and Hunter Lovins. *Natural Capitalism: Creating the Next Industrial Revolution.* Boston: Little, Brown, 1999.

Hayden, Dolores. *Building Suburbia: Green Fields and Urban Growth, 1820–2000.* New York: Vintage Books, 2004.

Hayes, Denis. *Rays of Hope: The Transition to a Post-Petroleum World.* New York: W. W. Norton, 1977.

Hays, Samuel P. *Beauty, Health, and Permanence: Environmental Politics in the United States, 1955–1985.* Cambridge: Cambridge University Press, 1987.

Heilbroner, Robert L. *An Inquiry into the Human Prospect.* New York: Norton, 1974.

———. *The Worldly Philosophers: The Lives, Times, and Ideas of the Great Economic Thinkers.* New York: Simon and Schuster, 1953.

Heinberg, Richard. *The End of Growth: Adapting to Our New Economic Reality.* Gabriola, BC: New Society, 2011.

———. *The Party's Over: Oil, War and the Fate of Industrial Societies.* Gabriola, BC: New Society, 2003.

———. *Peak Everything: Waking Up to the Century of Declines.* Gabriola, BC: New Society, 2007.

Helmreich, William B. *The Enduring Community: The Jews of Newark and MetroWest.* New Brunswick, NJ: Transaction, 1999.

Helvarg, David. *The War Against the Greens: The Wise-Use Movement, the New Right and Anti-Environmental Violence.* San Francisco: Sierra Club Books, 1994.

Herrick, Charles N. "Objectivity Versus Narrative Coherence: Science, Environmental Policy, and the U.S. Data Quality Act." *Environmental Science and Policy* 7, no. 5 (2004): 419–433.

Himelfarb, Richard, and Rosanna Perotti, eds. *Principle over Politics? The Domestic Policy of the George H. W. Bush Presidency.* Westport, CT: Praeger, 2004.

Hirsch, Fred. *Social Limits to Growth.* Cambridge, MA: Harvard University Press, 1976.

Hoff, Derek Seabury. "'Kick That Population Commission in the Ass': The Nixon Administration, the Commission on Population Growth and the American Future, and the Defusing of the Population Bomb." *Journal of Policy History* 22, no. 1 (2010): 23–63.

————. *The State and the Stork: The Population Debate and Policy Making in US History.* Chicago: University of Chicago Press, 2012.

Hoffman, Andrew J. *From Heresy to Dogma: An Institutional History of Corporate Environmentalism.* Expanded ed. Stanford, CA: Stanford University Press, 2001.

Holden, Constance. "Ehrlich Versus Commoner: An Environmental Fallout." *Science* 177 (1972): 245–247.

————. "Simon and Kahn Versus Global 2000." *Science* 221 (1983): 341–343.

Holdren, John P., and Paul R. Ehrlich, eds. *Global Ecology: Readings Toward a Rational Strategy for Man.* New York: Harcourt Brace Jovanovich, 1971.

Hollingsworth, Peggie J., ed. *Unfettered Expression: Freedom in American Intellectual Life.* Ann Arbor: University of Michigan Press, 2000.

Huber, Peter. *Hard Green: Saving the Environment from the Environmentalists: A Conservative Manifesto.* New York: Basic Books, 1999.

Hume, David. *An Enquiry Concerning the Principles of Morals* [1777]. Project Gutenberg Ebook #4320 [1912 reprint ed.].

Jackson, Kenneth T. *Crabgrass Frontier: The Suburbanization of the United States.* New York: Oxford University Press, 1985.

Jacobs, Meg, William J. Novak, and Julian E. Zelizer, eds. *The Democratic Experiment: New Directions in American Political History.* Princeton, NJ: Princeton University Press, 2003.

Jacoby, Karl. *Crimes Against Nature: Squatters, Poachers, Thieves, and the Hidden History of American Conservation.* Berkeley: University of California Press, 2001.

James, William. *Pragmatism.* New York: Longmans, Green, 1907.

Jarrett, H., ed. *Environmental Quality in a Growing Economy.* Baltimore, MD: Johns Hopkins University Press, 1966.

Johnson, M. P., A. D. Keith, and Paul R. Ehrlich. "The Population Biology of the Butterfly, Euphydryas Editha VII. Has E. Editha Evolved a Serpentine Race?" *Evolution* 22, no. 2 (1968): 422–423.

Kahn, Herman. *The Coming Boom: Economic, Political, and Social.* New York: Simon and Schuster, 1982.

———. *On Thermonuclear War.* Princeton, NJ: Princeton University Press, 1960.

Kahn, Herman, and Ernest Schneider. "Globaloney 2000." *Policy Review* 16 (Spring 1981): 129–147.

Kahn, Herman, and Anthony J. Wiener. "The Next Thirty-Three Years: A Framework for Speculation." *Daedalus* 96, no. 3 (1967): 705–732.

Kalman, Laura. *Right Star Rising: A New Politics, 1974–1980.* New York: W. W. Norton, 2010.

Kaysen, Carl. "The Computer That Printed Out W*O*L*F*." *Foreign Affairs* 50, no. 4 (1972): 660–668.

Kennedy, Robert Francis. *Crimes Against Nature: How George W. Bush and His Corporate Pals Are Plundering the Country and Hijacking Our Democracy.* New York: HarperCollins, 2004.

Kenny, Charles. *Getting Better: Why Global Development Is Succeeding—and How We Can Improve the World Even More.* New York: Basic Books, 2011.

Kevles, Daniel J. *In the Name of Eugenics: Genetics and the Uses of Human Heredity.* New York: Knopf, 1985.

Kiel, Katherine, Victor Matheson, and Kevin Golembiewski. "Luck or Skill? An Examination of the Ehrlich-Simon Bet." *Ecological Economics* 69 (2010): 1365–1367.

Kingsland, Sharon E. *The Evolution of American Ecology, 1890–2000.* Baltimore: Johns Hopkins University Press, 2005.

———. *Modeling Nature: Episodes in the History of Population Ecology.* 2nd ed. Chicago: University of Chicago Press, 1995.

Kirk, Andrew G. *Counterculture Green: The Whole Earth Catalog and American Environmentalism.* Lawrence: University Press of Kansas, 2007.

Klare, Michael T. *The Race for What's Left: The Global Scramble for the World's Last Resources.* New York: Metropolitan Books, 2012.

Kozinski, Alex. "Review: Gore Wars." *Michigan Law Review* 100, no. 6 (2002): 1742–1767.

Kunstler, James Howard. *The Long Emergency: Surviving the Converging Catastrophes of the Twenty-First Century.* New York: Atlantic Monthly Press, 2005.

Kuznets, Simon. "Population and Economic Growth." *Proceedings of the American Philosophical Society* 111, no. 3 (1967): 170–193.

Lader, Lawrence. *Breeding Ourselves to Death.* New York: Ballantine, 1971.

Lash, Jonathan, Katherine Gillman, and David Sheridan. *A Season of Spoils: The Story of the Reagan Administration's Attack on the Environment.* New York: Pantheon Books, 1984.

Lazarus, Richard J. *The Making of Environmental Law.* Chicago: University of Chicago Press, 2004.

Lewis, Michael. *The Big Short: Inside the Doomsday Machine.* New York: W. W. Norton, 2010.

Lilienfeld, Robert. *The Rise of Systems Theory: An Ideological Analysis.* New York: Wiley, 1978.

Lomborg, Bjørn. *The Skeptical Environmentalist: Measuring the Real State of the World.* Cambridge: Cambridge University Press, 2001.

Lovins, Amory B. *Soft Energy Paths: Toward a Durable Peace.* San Francisco: Friends of the Earth International, 1977.

Maddox, John Royden. *The Doomsday Syndrome.* London: Macmillan, 1972.

Malakoff, David. "Are More People Necessarily a Problem?" *Science* 333 (2011): 544–546.

Malthus, Thomas R. *An Essay on the Principle of Population as It Affects the Future Improvement of Society.* London: J. Johnson, 1798.

Mann, Michael E. *The Hockey Stick and the Climate Wars: Dispatches from the Front Lines.* New York: Columbia University Press, 2012.

Manne, Alan S. "Waiting for the Breeder." *Review of Economic Studies* 41 (1974): 47–65.

Mason, Larry G., Paul R. Ehrlich, and Thomas C. Emmel. "The Population Biology of the Butterfly, Euphydryas Editha. V. Character Clusters and Asymmetry." *Evolution* 21, no. 1 (1967): 85–91.

Mattson, Kevin. *What the Heck Are You Up to, Mr. President? Jimmy Carter, America's Malaise, and the Speech That Should Have Changed the Country.* New York: Bloomsbury, 2009.

McAnany, Patricia Ann, and Norman Yoffee, eds. *Questioning Collapse: Human Resilience, Ecological Vulnerability, and the Aftermath of Empire.* New York: Cambridge University Press, 2010.

McBay, Aric. *Peak Oil Survival: Preparation for Life After Gridcrash.* Guilford, CT: Lyons Press, 2006.

McCloskey, J. Michael. *In the Thick of It: My Life in the Sierra Club.* Washington, DC: Shearwater Books, 2005.

McCommons, James. *Waiting on a Train: The Embattled Future of Passenger Rail Service.* White River Junction, VT: Chelsea Green, 2009.

McEvoy, Arthur F. *The Fisherman's Problem: Ecology and Law in the California Fisheries, 1850–1980.* Cambridge: Cambridge University Press, 1986.

McKibben, Bill. *Eaarth: Making a Life on a Tough New Planet.* New York: Times Books, 2010.

———. *Maybe One: A Personal and Environmental Argument for Single-Child Families.* New York: Simon and Schuster, 1998.

McNeill, J. R. *Something New Under the Sun: An Environmental History of the Twentieth-Century World.* New York: W. W. Norton, 2000.

Meadows, Dennis L., et al. *Beyond Growth: Essays on Alternative Futures,* ed. William R. Burch, and F. Herbert Bormann. Yale University, School of

Forestry and Environmental Studies, Bulletin no. 88. New Haven: Yale
   University, 1975.
———. *The Limits to Growth.* 2nd ed. New York: Universe Books, 1974.
Meadows, Donella H., Dennis L. Meadows, and Jørgen Randers. *Beyond the
   Limits: Confronting Global Collapse, Envisioning a Sustainable Future.*
   Post Mills, VT: Chelsea Green, 1992.
Meadows, Donella H., John M. Richardson, and Gerhart Bruckmann.
   Groping in the Dark: The First Decade of Global Modelling. Chichester,
   UK: Wiley, 1982.
Medhurst, Martin J., ed. *The Rhetorical Presidency of George H. W. Bush.*
   College Station: Texas A&M University Press, 2006.
Meek, Ronald L., ed. *Marx and Engels on Malthus: Selections from the Writings
   of Marx and Engels Dealing with the Theories of Thomas Robert Malthus.*
   New York: International Publishers, 1954.
Milazzo, Paul Charles. *Unlikely Environmentalists: Congress and Clean Water,
   1945–1972.* Lawrence: University Press of Kansas, 2006.
Miller, Timothy. "The Sixties-Era Communes." In *Imagine Nation: The
   American Counterculture of the 1960s and 1970s,* ed. Peter Braunstein
   and Michael William Doyle, 327–351. New York: Routledge, 2002.
Mondale, Walter F., and Dave Hage. *The Good Fight: A Life in Liberal Politics.*
   New York: Scribner, 2010.
Montagu, Ashley, ed. *The Concept of Race.* New York: Free Press of Glencoe,
   1964.
Mooney, Chris. *The Republican War on Science.* New York: Basic Books, 2005.
Moore, Stephen. "Julian Simon Remembered: It's a Wonderful Life." *Cato
   Policy Report,* March/April 1998.
Moore, Stephen, and Julian L. Simon. *It's Getting Better All the Time: 100
   Greatest Trends of the Last 100 Years.* Washington, DC: Cato Institute,
   2000.
Myers, Norman, and Julian L. Simon. *Scarcity or Abundance? A Debate on the
   Environment.* New York: W. W. Norton, 1994.
National Academy of Sciences. *Rapid Population Growth: Consequences and
   Policy Implications.* Baltimore: Johns Hopkins University Press, 1971.
National Research Council. *Population Growth and Economic Development:
   Policy Questions.* Washington, DC: National Academies Press, 1986.
Nordhaus, William D. "Resources as a Constraint on Growth." *American
   Economic Review* 64, no. 2 (1974): 22–26.
———. "World Dynamics: Measurement Without Data." *Economic Journal* 83
   (1973): 1156–1183.
Nordhaus, William D., Hendrik Houthakker, and Robert Solow. "The
   Allocation of Energy Resources." *Brookings Papers on Economic Activity*
   1973, no. 3 (1973): 529–576.

Nordhaus, William D., Robert N. Stavins, and Martin L. Weitzman. "Lethal Model 2: The Limits to Growth Revisited." *Brookings Papers on Economic Activity* 1992, no. 2 (1992): 1–59.

Norgaard, Richard B. "Scarcity and Growth: How Does It Look Today?" *American Journal of Agricultural Economics* 57, no. 5 (1975): 810–814.

Olson, Steve. "Knowing How to Pick a Fight." *Seed Magazine*, 4 August 2009.

Oreskes, Naomi. "Science and Public Policy: What's Proof Got to Do with It?" *Environmental Science and Policy* 7, no. 5 (2004): 369–383.

Oreskes, Naomi, and Erik M. Conway. *Merchants of Doubt: How a Handful of Scientists Obscured the Truth on Issues from Tobacco Smoke to Global Warming*. New York: Bloomsbury, 2011.

Orr, David W. "The Labors of Sisyphus." *Conservation Biology* 16, no. 4 (2002): 857–860.

Osborn, Fairfield. *Our Crowded Planet: Essays on the Pressures of Population*. Garden City, NY: Doubleday, 1962.

———. *Our Plundered Planet*. Boston: Little, Brown, 1948.

Paddock, William, and Paul Paddock. *Famine, 1975! America's Decision: Who Will Survive?* Boston: Little, Brown, 1967.

Passell, Peter, and Leonard Ross. *The Retreat from Riches: Affluence and Its Enemies*. New York: Viking, 1973.

Paul Ehrlich and the Population Bomb. Princeton, NJ: Films for the Humanities and Sciences, 1996.

Phillips-Fein, Kim. *Invisible Hands: The Making of the Conservative Movement from the New Deal to Reagan*. New York: W. W. Norton, 2010.

Pielke, Roger A., Jr. *The Honest Broker: Making Sense of Science in Policy and Politics*. Cambridge: Cambridge University Press, 2007.

———. "When Scientists Politicize Science: Making Sense of Controversy over the Skeptical Environmentalist." *Environmental Science and Policy* 7, no. 5 (2004): 405–417.

Pilzer, Paul Zane. *God Wants You to Be Rich*. New York: Fireside, 1997.

Pimm, Stuart, and Jeff Harvey. "No Need to Worry About the Future." *Nature* 414 (2001): 149–150.

Pirages, Dennis, and Ken Cousins, eds. *From Resource Scarcity to Ecological Security: Exploring New Limits to Growth*. Cambridge, MA: MIT Press, 2005.

Pirages, Dennis, and Paul R. Ehrlich. *Ark II: Social Response to Environmental Imperatives*. San Francisco: W. H. Freeman, 1973.

"The Plowboy Interview: Paul Ehrlich." *Mother Earth News*, July–August 1974.

Pohlman, Edward, ed. *Population: A Clash of Prophets*. New York: New American Library, 1973.

Pooley, Eric. *The Climate War: True Believers, Power Brokers, and the Fight to Save the Earth*. New York: Hyperion, 2010.

Posner, Richard A. *Public Intellectuals: A Study of Decline; with a New Preface and Epilogue*. Cambridge, MA: Harvard University Press, 2003.

Price, David. "Carrying Capacity Reconsidered." *Population and Environment* 21, no. 1 (1999): 5–26.

Rathlesberger, James, ed. *Nixon and the Environment; the Politics of Devastation: Thirteen Essays*. A League of Conservation Voters Report. New York: Taurus Communications, 1972.

Ray, Dixy Lee, and Louis R. Guzzo. *Trashing the Planet: How Science Can Help Us Deal with Acid Rain, Depletion of the Ozone, and Nuclear Waste*. Washington, DC: Regnery, 1990.

Reagan, Ronald. *An American Life*. New York: Simon and Schuster, 1990.

———. *The Reagan Diaries*. New York: HarperCollins, 2007.

———. *Reagan, in His Own Hand: The Writings of Ronald Reagan That Reveal His Revolutionary Vision for America*. New York: Free Press, 2001.

Real, Leslie, and James H. Brown, eds. *Foundations of Ecology: Classic Papers with Commentaries*. Chicago: University of Chicago Press, 1991.

Regis, Ed. "The Doomslayer." *Wired* 5, no. 2 (1997): 136.

Reinhardt, Forest L. *Down to Earth: Applying Business Principles to Environmental Management*. Boston: Harvard Business School Press, 2000.

Revelle, Roger. "Review: Paul Ehrlich: New High Priest of Ecocatastrophe." *Family Planning Perspectives* 3, no. 2 (1971): 66–70.

Roberts, Godfrey, ed. *Population Policy: Contemporary Issues*. New York: Praeger, 1990.

Robertson, Thomas. *The Malthusian Moment: Global Population Growth and the Birth of American Environmentalism*. New Brunswick, NJ: Rutgers University Press, 2012.

Rodgers, Daniel T. *Age of Fracture*. Cambridge, MA: Belknap Press of Harvard University Press, 2011.

Rome, Adam. *The Bulldozer in the Countryside: Suburban Sprawl and the Rise of American Environmentalism*. New York: Cambridge University Press, 2001.

———. "'Give Earth a Chance': The Environmental Movement and the Sixties." *Journal of American History* 90, no. 2 (2003): 525–554.

Rorabaugh, W. J. *Berkeley at War: The 1960s*. New York: Oxford University Press, 1989.

Ross, Andrew. *Strange Weather: Culture, Science, and Technology in the Age of Limits*. London: Verso, 1991.

Rothman, Hal K. *The Greening of a Nation?: Environmentalism in the U.S. Since 1945*. Belmont, CA: Wadsworth, 1997.

Rotter, Andrew. "Empires of the Senses: How Seeing, Hearing, Smelling, Tasting, and Touching Shaped Imperial Encounters." *Diplomatic History* 35, no. 1 (2011): 3–19.

Rowell, Andrew. *Green Backlash: Global Subversion of the Environmental Movement.* London: Routledge, 1996.

Rubin, Jeff. *The End of Growth.* Toronto: Random House Canada, 2012.

Russell, Edmund. *War and Nature: Fighting Humans and Insects with Chemicals from World War I to Silent Spring.* New York: Cambridge University Press, 2001.

Sabin, Paul. "Crisis and Continuity in U.S. Oil Politics, 1965–1980." *Journal of American History* 99, no. 1 (2012): 177–186.

———. *Crude Politics: The California Oil Market, 1900–1940.* Berkeley: University of California Press, 2005.

———. "Searching for Middle Ground: Native Communities and Oil Extraction in the Northern and Central Ecuadorian Amazon, 1967–1993." *Environmental History* 3, no. 2 (1998): 144–168.

———. "'The Ultimate Environmental Dilemma': Making a Place for Historians in the Climate Change and Energy Debates." *Environmental History* 15, no. 1 (2010): 76–93.

———. "Voices from the Hydrocarbon Frontier: Canada's Mackenzie Valley Pipeline Inquiry (1974–1977)." *Environmental History Review* 19, no. 1 (1995): 17–48.

Salmon, Jack D. "Politics of Scarcity Versus Technological Optimism: A Possible Reconciliation?" *International Studies Quarterly* 21, no. 4 (1977): 701–720.

Sarewitz, Daniel. "How Science Makes Environmental Controversies Worse." *Environmental Science and Policy* 7, no. 5 (2004): 385–403.

Sayre, Nathan F. "The Genesis, History, and Limits of Carrying Capacity." *Annals of the Association of American Geographers* 98, no. 1 (2008): 120–134.

Schneider, Stephen H. *Science as a Contact Sport: Inside the Battle to Save Earth's Climate.* Washington, DC: National Geographic, 2009.

Schulman, Bruce J. *The Seventies: The Great Shift in American Culture, Society, and Politics.* New York: Free Press, 2001.

Schulman, Bruce, and Julian Zelizer, eds. *Rightward Bound: Making America Conservative in the 1970s.* Cambridge, MA: Harvard University Press, 2008.

Schumacher, E. F. *Small Is Beautiful: Economics as If People Mattered.* New York: Harper and Row, 1973.

Sen, Amartya. *Poverty and Famines: An Essay on Entitlement and Deprivation.* New York: Oxford University Press, 1981.

Shabecoff, Philip. *A Fierce Green Fire: The American Environmental Movement.* New York: Hill and Wang, 1993.

Shanley, Robert A. *Presidential Influence and Environmental Policy.* Westport, CT: Greenwood, 1992.

Short, C. Brant. *Ronald Reagan and the Public Lands: America's Conservation Debate, 1979–1984.* College Station: Texas A&M University Press, 1989.

Simmons, Matthew. *Twilight in the Desert: The Coming Saudi Oil Shock and the World Economy.* New York: Wiley, 2005.

Simon, Julian L. "Airline Overbooking: A Rejoinder." *Journal of Transport Economics and Policy* 4, no. 2 (1970): 212–213.

———. "Airline Overbooking: The State of the Art: A Reply." *Journal of Transport Economics and Policy* 6, no. 3 (September 1972): 254–256.

———. "The Airline Oversales Auction Plan: The Results." *Journal of Transport Economics and Policy* 28, no. 3 (1994): 319–323.

———. "An Almost Practical Solution to Airline Overbooking." *Journal of Transport Economics and Policy* 2, no. 2 (1968): 201–202.

———. "Basic Data Concerning Immigration into the United States." *Annals of the American Academy of Political and Social Science* 487 (September 1986): 12–56.

———. *Basic Research Methods in Social Science: The Art of Empirical Investigation.* 2nd ed. New York: Random House, 1978.

———. *Developing Decision-Making Skills for Business.* Armonk, NY: M. E. Sharpe, 2001.

———. "Does Doom Loom?" *Reason,* April 1984, 34.

———. "Does Economic Growth Imply a Growth in Welfare?" *Journal of Economic Issues* 7, no. 1 (1973): 130–136.

———. *The Economic Consequences of Immigration.* Oxford: Basil Blackwell, 1989.

———. *Economics Against the Grain.* Brookfield, VT: Edward Elgar, 1998.

———. *The Economics of Population: Classic Writings.* New Brunswick, NJ: Transaction, 1998.

———. *The Economics of Population Growth.* Princeton, NJ: Princeton University Press, 1977.

———. "Economic Thought About Population Consequences: Some Reflections." *Journal of Population Economics* 6, no. 2 (1993): 137–152.

———. "The Effect of Income on Fertility." *Population Studies* 23, no. 3 (1969): 327–341.

———. *The Effects of Income on Fertility.* Chapel Hill, NC: Carolina Population Center, 1974.

———. "The Effects of Population on Nutrition and Economic Well-Being." *Journal of Interdisciplinary History* 14, no. 2 (1983): 413–437.

———. "Environmental Disruption or Environmental Improvement?" *Social Science Quarterly* 62, no. 1 (1981): 30–43.

———. "Family Planning Prospects in Less-Developed Countries, and a Cost-Benefit Analysis of Various Alternatives." *Economic Journal* 80 (1970): 58–71.

————. "Global Confusion 1980: A Hard Look at the Global 2000 Report." *Public Interest,* 3 March 1981, 3–20.

————. *Good Mood: The New Psychology of Overcoming Depression.* Chicago: Open Court, 1993.

————. *The Great Breakthrough and Its Cause.* Ann Arbor: University of Michigan Press, 2000.

————. *Hoodwinking the Nation.* New Brunswick, NJ: Transaction, 1999.

————. *How Do Immigrants Affect Us Economically?* Washington, DC: Georgetown University, Center for Immigration Policy and Refugee Assistance, 1985.

————. *How to Start and Operate a Mail-Order Business.* 4th ed. New York: McGraw-Hill, 1987.

————. "A Huge Marketing Research Task: Birth Control." *Journal of Marketing Research* 5, no. 1 (1968): 21–27.

————. "Immigrants, Taxes, and Welfare in the United States." *Population and Development Review* 10, no. 1 (1984): 55–69.

————. *Issues in the Economics of Advertising.* Urbana: University of Illinois Press, 1970.

————. *A Life Against the Grain: The Autobiography of an Unconventional Economist.* New Brunswick, NJ: Transaction, 2002.

————. "The Mixed Effects of Income Upon Successive Births May Explain the Convergence Phenomenon." *Population Studies* 29, no. 1 (1975): 109–122.

————. "Paul Ehrlich Saying It Is So Doesn't Make It So," *Social Science Quarterly* 63, no. 2 (1982): 381.

————. "The Per-Capita-Income Criterion and Natality Policies in Poor Countries." *Demography* 7, no. 3 (1970): 369–378.

————. *Population and Development in Poor Countries: Selected Essays.* Princeton, NJ: Princeton University Press, 1992.

————. "Population Growth May Be Good for LDCs in the Long Run: A Richer Simulation Model." *Economic Development and Cultural Change* 24, no. 2 (1976): 309–337.

————. *Population Matters: People, Resources, Environment, and Immigration.* New Brunswick, NJ: Transaction, 1990.

————. "The Positive Effect of Population Growth on Agricultural Saving in Irrigation Systems." *Review of Economics and Statistics* 57, no. 1 (1975): 71–79.

————. "Puzzles and Further Explorations in the Interrelationships of Successive Births with Husband's Income, Spouse's Education and Race." *Demography* 12, no. 2 (1975): 259–274.

————. *Research in Population Economics.* Greenwich, CT: JAI Press, 1978.

————. "Resources, Population, Environment: An Oversupply of False Bad News." *Science,* 27 June 1980, 1431–1437.

————. "The Role of Bonuses and Persuasive Propaganda in the Reduction of Birth Rates." *Economic Development and Cultural Change* 16, no. 3 (1968): 404–411.

————. "The Scarcity of Raw Materials." *Atlantic Monthly*, June 1981, 33–41.

————. "A Scheme to Improve Air Travel." *Journal of Policy Analysis and Management* 2, no. 3 (1983): 465–466.

————. "Some 'Marketing Correct' Recommendations for Family Planning Campaigns." *Demography* 5, no. 1 (1968): 504–507.

————. *Theory of Population and Economic Growth.* Oxford: Blackwell, 1986.

————. "There Is No Low-Level Fertility and Development Trap." *Population Studies* 34, no. 3 (1980): 476–486.

————. *The Ultimate Resource.* Princeton, NJ: Princeton University Press, 1981.

————. *The Ultimate Resource 2.* Rev. ed. Princeton, NJ: Princeton University Press, 1996.

————. "The Unreported Revolution in Population Economics." *Public Interest* 101 (Fall 1990): 89–100.

————. "The Value of Avoided Births to Underdeveloped Countries." *Population Studies* 23, no. 1 (1969): 61–68.

————. "The Welfare Effect on an Additional Child Cannot Be Stated Simply and Unequivocally." *Demography* 12, no. 1 (1975): 89–105.

————. "World Food Supplies." *Atlantic Monthly*, July 1981, 72–76.

————. "World Population Growth." *Atlantic Monthly*, August 1981, 70–76.

————. "The Worth Today of United States Slaves' Imputed Wages." *Journal of Economic Issues* 5, no. 3 (1971): 110–113.

Simon, Julian L., and William F. Buckley, Jr. "Answer to Malthus? Julian Simon Interviewed by William Buckley." *Population and Development Review* 8, no. 1 (1982): 205–218.

Simon, Julian L., and David M. Gardner. "World Food Needs and 'New Proteins.'" *Economic Development and Cultural Change* 17, no. 4 (1969): 520–526.

Simon, Julian L., and Herman Kahn, eds. *The Resourceful Earth: A Response to Global 2000.* Oxford: Basil Blackwell, 1984.

Simon, Julian L., and G. Visvabhanathy. "The Auction Solution to Airline Overbooking: The Data Fit the Theory." *Journal of Transport Economics and Policy* 11, no. 3 (1977): 277–283.

Simon, Julian L., E. Calvin Beisner, and John Phelps. *The State of Humanity.* Oxford: Blackwell 1995.

Simon, Rita J. "American Women and Crime." *Annals of the American Academy of Political and Social Science* 423 (January 1976): 31–46.

————. "Black Attitudes Toward Transracial Adoption." *Phylon* 39, no. 2 (1978): 135–142.

———. *A Comparative Perspective on Major Social Problems.* Lanham, MD: Lexington Books, 2001.

———. *Immigration and American Public Policy.* Beverly Hills, CA: Sage, 1986.

———. *The Jury: Its Role in American Society.* Lexington, MA: D. C. Heath, 1980.

———. "Public Attitudes Toward Population and Pollution." *Public Opinion Quarterly* 35, no. 1 (1971): 93–99.

———. [Geraldine R. Muntz, pseud.] "Some Observations on the Function of Women Sociologists at Sociology Conventions." *American Sociologist* 2, no. 3 (1967): 158–159.

———. "A Survey of Faculty Attitudes Toward a Labor Dispute at Their University." *AAUP Bulletin* 52, no. 2 (1966): 223–224.

———. *Transracial Adoptees and Their Families: A Study of Identity and Commitment.* New York: Praeger, 1987.

Simon, Rita J., ed. *As We Saw the Thirties: Essays on Social and Political Movements of a Decade.* Urbana: University of Illinois Press, 1967.

Simon, Rita J., and Wendell Shackelford. "The Defense of Insanity: A Survey of Legal and Psychiatric Opinion." *Public Opinion Quarterly* 29, no. 3 (1965): 411–424.

Simon, Rita J., and Julian L. Simon. "The Jewish Dimension Among Recent Soviet Immigrants to the United States." *Jewish Social Studies* 44, no. 3/4 (1982): 283–290.

Simon, Rita J., Shirley Clark, and Kathleen Galway. "The Woman Ph.D.—A Recent Profile." *Social Problems,* Autumn 1967, 221–236.

Simon, Rita J., Shirley M. Clark, and Larry L. Tifft. "Of Nepotism, Marriage, and the Pursuit of an Academic Career." *Sociology of Education* 39, no. 4 (1966): 344–358.

Simpson, Ralph David, Michael A. Toman, and Robert U. Ayres, eds. *Scarcity and Growth Revisited: Natural Resources and the Environment in the New Millennium.* Washington, DC: Resources for the Future, 2005.

Singer, Steve, et al. "Bad News: Is It True?" *Science* 210 (1980): 1296–1308.

Smil, Vaclav. *Global Catastrophes and Trends: The Next Fifty Years.* Cambridge, MA: MIT Press, 2008.

———. "Limits to Growth Revisited: A Review Essay." *Population and Development Review* 31, no. 1 (2005): 157–164.

Smith, Kirk R., Fereidun Fesharaki, and John P. Holdren, eds. *Earth and the Human Future: Essays in Honor of Harrison Brown.* Boulder, CO: Westview, 1986.

Smith, V. Kerry, ed. *Scarcity and Growth Reconsidered.* Baltimore: Resources for the Future, 1979.

Snow, C. P. *The State of Siege.* New York: Scribner, 1969.

Solow, Robert M. "The Economics of Resources or the Resources of Economics." *American Economic Review* 64, no. 2 (1974): 1–14.

———. "The Economist's Approach to Pollution and Its Control." *Science* 173 (1971): 498–503.

———. "Intergenerational Equity and Exhaustible Resources." *Review of Economic Studies* 41 (1974): 29–45.

———. "Is the End of the World at Hand?" *Challenge* 16, no. 1 (1973): 39–50.

———. "Notes on 'Doomsday Models.'" *Proceedings of the National Academy of Sciences of the United States of America* 69 (1972): 3832–3833.

Solow, Robert M., and Frederic Y. Wan. "Extraction Costs in the Theory of Exhaustible Resources." *Bell Journal of Economics* 7, no. 2 (1976): 359–370.

Speth, James Gustave. *The Bridge at the Edge of the World: Capitalism, the Environment, and Crossing from Crisis to Sustainability.* New Haven: Yale University Press, 2008.

———. *Red Sky at Morning: America and the Crisis of the Global Environment.* New Haven: Yale University Press, 2004.

Stacks, John F. *Watershed: The Campaign for the Presidency, 1980.* New York: Times Books, 1981.

Steffen, Will, et al. *Global Change and the Earth System: A Planet Under Pressure.* Berlin: Springer, 2004.

Stiglitz, Joseph E. "Growth with Exhaustible Natural Resources: Efficient and Optimal Growth Paths." *Review of Economic Studies* 41 (1974): 123–137.

———. "Growth with Exhaustible Natural Resources: The Competitive Economy." *Review of Economic Studies* 41 (1974): 139–152.

Stradling, David, and Richard Stradling. "Perceptions of the Burning River: Deindustrialization and Cleveland's Cuyahoga River." *Environmental History* 13 (July 2008): 515–535.

Sunstein, Cass R. *Risk and Reason: Safety, Law, and the Environment.* Cambridge: Cambridge University Press, 2002.

Swaney, James A. "Julian Simon Versus the Ehrlichs: An Institutionalist Perspective." *Journal of Economic Issues* 25, no. 2 (1991): 499–509.

Teitelbaum, Michael S. *A Question of Numbers: High Migration, Low Fertility, and the Politics of National Identity.* New York: Hill and Wang, 1998.

Teitelbaum, Michael S., and J. M. Winter, eds. *The Fear of Population Decline.* Orlando, FL: Academic Press, 1985.

———. *Population and Resources in Western Intellectual Traditions.* Cambridge: Cambridge University Press, 1989.

Teles, Steven M. *The Rise of the Conservative Legal Movement: The Battle for Control of the Law.* Princeton, NJ: Princeton University Press, 2010.

Theobald, Robert. *Habit and Habitat.* Englewood Cliffs, NJ: Prentice-Hall, 1972.

Thoburn, John T. "The Tin Industry Since the Collapse of the International
    Tin Agreement." *Resources Policy* 20, no. 2 (1994): 125–133.
———. *Tin in the World Economy*. Edinburgh: Edinburgh University Press,
    1994.
Tierney, John. "Betting on the Planet." *New York Times Magazine*, 2 December
    1990, 81.
Tobey, Ronald C. *Saving the Prairies: The Life Cycle of the Founding School of
    American Plant Ecology, 1895–1955*. Berkeley: University of California
    Press, 1981.
Tol, Richard S. J. "The Economic Effects of Climate Change." *Journal of
    Economic Perspectives* 23, no. 2 (2009): 29–51.
Turner, Tom. "The Vindication of a Public Scholar." *Earth Island Journal*
    (Summer 2009): 50–56.
US Congress. House. Subcommittee on Foreign Economic Policy. "Foreign
    Policy Implications of the Energy Crisis." 92nd Cong., 2nd sess., 21, 26,
    27 September, 3 October 1972.
US Congress. Senate. Committee on Commerce and Committee on
    Government Operations. "Domestic Supply Information Act: Joint
    Hearings before the Committee on Commerce and Committee on
    Government Operations." 93rd Cong., 2nd Sess., 29 April, 9, 10 May,
    17 June 1974.
US Council on Environmental Quality and Department of State. *The Global
    2000 Report to the President of the U.S.: Entering the 21st Century*. Vol. 1:
    The Summary Report. New York: Pergamon, 1980.
US Environmental Protection Agency. Office of Air and Radiation. *The
    Benefits and Costs of the Clean Air Act from 1990 to 2020: Final Report*.
    Washington, DC: US Environmental Protection Agency, 2011.
Vietor, Richard H. K. *Contrived Competition: Regulation and Deregulation in
    America*. Cambridge, MA: Belknap Press of Harvard University Press,
    1994.
———. *Energy Policy in America Since 1945: A Study of Business Government
    Relations*. Cambridge: Cambridge University Press, 1984.
Vig, Norman J., and Michael E. Kraft, eds. *Environmental Policy in the 1990s:
    Toward a New Agenda*. 2nd ed. Washington, DC: CQ Press, 1994.
Vogt, William. *Road to Survival*. New York: W. Sloane, 1948.
Wade, Green. "The Militant Malthusians." *Saturday Review*, 11 March 1972, 49.
Wallace, Alfred Russel. *My Life: A Record of Events and Opinions*. Vol. 1. New
    York: Dodd, Mead, 1905.
Wang, Haidong, et al. "Age-Specific and Sex-Specific Mortality in 187
    Countries, 1970–2010: A Systematic Analysis for the Global Burden
    of Disease Study 2010." *Lancet* 380 (2012): 2071–2094.
Wargo, John. *Green Intelligence: Creating Environments That Protect Human
    Health*. New Haven: Yale University Press, 2009.

Warren, Louis S. *The Hunter's Game: Poachers and Conservationists in Twentieth-Century America.* New Haven, CT: Yale University Press, 1997.

Watt, James G., and Doug Wead. *The Courage of a Conservative.* New York: Simon and Schuster, 1985.

Wattenberg, Ben. *The Birth Dearth.* New York: Pharos Books, 1987.

———. *The Good News Is the Bad News Is Wrong.* New York: Simon and Schuster, 1984.

Weiss, Harvey, and Raymond S. Bradley. "What Drives Societal Collapse?" *Science* 291 (2001): 609–610.

Weitzman, Martin L. "Prices Vs. Quantities." *Review of Economic Studies* 41, no. 4 (1974): 477–491.

Wellock, Thomas R. *Preserving the Nation: The Conservation and Environmental Movements, 1870–2000.* Wheeling, IL: Harlan Davidson, 2007.

Whitaker, John C. *Striking a Balance: Environment and Natural Resources Policy in the Nixon-Ford Years.* Washington, DC: American Enterprise Institute for Public Policy Research, 1976.

White, Richard. *Railroaded: The Transcontinentals and the Making of Modern America.* New York: W. W. Norton, 2011.

Wildavsky, Aaron B. *But Is It True? A Citizen's Guide to Environmental Health and Safety Issues.* Cambridge, MA: Harvard University Press, 1995.

Wilentz, Sean. *The Age of Reagan: A History, 1974–2008.* New York: Harper, 2008.

Williamson, Francis S. L. "Population Pollution." *BioScience* 19, no. 11 (1969): 979–983.

Windolf, Jim. "Sex, Drugs, and Soybeans." *Vanity Fair,* 5 April 2007.

Woolley, John T., and Gerhard Peters. *The American Presidency Project.* Online resource. Santa Barbara, CA. http://www.presidency.ucsb.edu.

Worster, Donald. *Nature's Economy: A History of Ecological Ideas.* 2nd ed. Cambridge: Cambridge University Press, 1994.

Worth, Kenneth D. *Peak Oil and the Second Great Depression (2010–2030): A Survival Guide for Investors and Savers After Peak Oil.* Denver, CO: Outskirts Press, 2010.

Zaretsky, Natasha. *No Direction Home: The American Family and the Fear of National Decline, 1968–1980.* Chapel Hill: University of North Carolina Press, 2007.

Zelizer, Julian E. *Jimmy Carter.* New York: Times Books, 2010.

———. "Reflections: Rethinking the History of American Conservatism." *Reviews in American History* 38, no. 2 (2010): 367–392.

Index

Page numbers in *italics* refer to illustrations.